PIMLICO

743

NATURE CURE

Richard Mabey's ground-breaking bestseller, *Flora Britannica*, won the British Book Awards' Illustrated Book of the Year and the Botanical Society of the British Isles' President's Award, and was runner-up for the BP Natural World Book Prize.

His previous books include *Food for Free*, *The Unofficial Countryside* and *The Common Ground*, as well as his intensely personal study of the nightingale, *Whistling in the Dark*. His biography of Gilbert White won the Whitbread Biography Award.

'*Nature Cure* is an assertion of the power of nature. It is Mabey's most personal book and also his best. It contains his most sustained and achieved thoughts on the natural world... If in the future we are to secure the survival of the natural world we need more than ever Richard Mabey's passionate and excited exploration of why it is worth saving not only for its own sake but also for ours.' Tim Dee, *Times Literary Supplement*

'Mabey's memoir on his recovery from crippling depression through a rediscovery of his love for nature is quite remarkable, both for its honesty and its total lack of ego. He has much to say on man's relationship with nature and the prose is both warm and fiercely intelligent.' Judges' verdict, Whitbread Book Awards, 2005

'Through his glorious use of language, he pursues strange and spontaneous quests through literature and the new countryside he is exploring... This book will help others see the world better as they walk through it.' Robin Hanbury-Tenison, *Country Life*

'Recovery was a slow but magical process: he fell in love, moved to East Anglia and picked up his pen again. This luminous memoir is the result, alive with vivid observations of nature's marvels, and the rediscovered realisation of how important it all is.' *Mail on Sunday*

'A beautifully observed book. It's hard to imagine Mabey's equal in writing about nature.' Caroline Gascoigne, Books of the Year, *Sunday Times*

'A rare treat, beautifully written.' Sir John Lister-Kaye, Books of the Year, *Scotsman*

'Profound and starkly beautiful.' *Bookseller*

'A marvellously hopeful read.' Jackie McGlone, *Herald*

'Mabey's forte is a quality of observation that combines a keen analytical edge with emotional passion and an artist's eye... [we experience] the pleasure of watching Mabey, tentatively at first, then more like the fledgling swift he flung in the air, resume the graceful intimacy with the natural world that he feared he had lost for good.' Jane Shilling, *Sunday Telegraph*

'Extraordinary power and authority [and] captivating beauty. It's not a self-help book; Mabey is not setting himself up for emulation. Instead we have an attempt to understand the forces that unite the human with the non-human, and what the relationship between them not only is, but, with patience and freedom from the more limiting obsessions of contemporary culture, can still become.' Paul Binding, *Independent on Sunday*

'Part love poem, part essay, part drama. Defying categorisation, it is on its own... *Nature Cure* is a compulsive read, a book to savour, to revisit, and to be thankful for.' James Robertson, *Natur Cymru*

'Mabey is one of our most eloquent and evocative nature writers. He depicts nature from a sense of supreme – non-religious – spirituality and with a strong emotional bond but devoid of sentimentality... he brings it alive as scarcely any other writer.' John Green, *Morning Star*

'A book about facing and adjusting to change, beautifully and sensitively written.' *Open Space*

'Part autobiography, part meditation on the relationship between nature and culture... Mabey understands that beautiful writing is a matter of never being bigger than your subject... and has not lost the childlike pleasure in nature that transports him and his readers to the gates of heaven.' Will Cohu, *Telegraph*

'Beautifully lyrical.' *Big Issue*

'This is a remarkable story of one man's recovery that offers hope for the human race.' Jane Anderson, *Radio Times*

'The story alone, of personal misfortune and emotional rescue, would have made this book compelling reading. What gives it even greater appeal is the interlaced account of his joyous voyage of discovery into the natural history of his new home patch, leading to a re-examination of a lifetime's attitudes about human relationships with landscape and its natural, and not-so-natural, inhabitants.' *BBC Wildlife*

'A rich, invigorating and deeply restorative broth of a book.' Paddy Woodworth, *Irish Times*

'I like to think that the house where I lived, with its creaking timbers and the sense of spiritual (if not actual) warmth that it conveys, also contributed its share of inspiration, bringing Richard back to life and providing a background to a book that offers, particularly to those who have been through personal earthquakes of their own, a story of hope and strength.' Terence Blacker (previous inhabitant of the house round which the book is based), *Independent*

NATURE CURE

RICHARD MABEY

PIMLICO

To Polly, with love

———

Published by Pimlico 2006

2 4 6 8 10 9 7 5 3 1

Copyright © Richard Mabey 2005

Anne Stevenson's poem 'Swifts', on page 162,
is published by Bloodaxe Books

First published in Great Britain in 2005 by
Chatto & Windus

Pimlico
Random House, 20 Vauxhall Bridge Road,
London SW1V 2SA

Random House Australia (Pty) Limited
20 Alfred Street, Milsons Point, Sydney,
New South Wales 2061, Australia

Random House New Zealand Limited
18 Poland Road, Glenfield,
Auckland 10, New Zealand

Random House (Pty) Limited
Isle of Houghton, Corner of Boundary Road & Carse O'Gowrie,
Houghton, 2198, South Africa

Random House UK Limited Reg. No. 954009

A CIP catalogue record for this book is available from the British Library

ISBN 9780701178321 (from Jan 2007)
ISBN 1844130967

Contents

Acknowledgements

I would like to express my heartfelt thanks to all those friends and helpers who, with patience, good humour and remarkable generosity, speeded my recovery, especially Di Brierley, Mike and Pooh Curtis, Roger Deakin, Mike Glasser, Francesca Greenoak, Tim Kidger, John Kilpatrick, the Royal Literary Fund, Caroline Soper, Justin Ward, Ian Wood. And to my sister Gill, for putting up with me as long as she did – and, of course, welcoming me back into the fold in the end.

My thanks, too, for trust and support that never wavered even in the bleakest times, to my agent, Vivien Green, and my publisher, Penny Hoare, who also helped me polish the final text. Roger Cazalet and Mark Cocker also generously read the typescript, and made invaluable and constructive criticisms.

Some short passages in the text appeared in a different form in the *Times*, the *Guardian* and *BBC Wildlife*. I owe a great debt to my editors there, Jane Wheatley, Annaleena Macafee and Rosamund Kidman Cox, for their encouragement, even for my most wayward ideas.

Finally, more thanks than I can express to Polly, who rescued me, took a chance, and with love and patience buoyed me up, and whose wisdom and sense of fun helped ease the moody storms that go with writing a book. This is her book as much as it is mine.

Flitting

I dwell on trifles like a child
I feel as ill becomes a man
And still my thoughts like weedlings wild
Grow up to blossom where they can.
                    John Clare, 'The Flitting'

IT'S OCTOBER, AN INDIAN summer. I'm standing on the threshold like some callow teenager, about to move house for the first time in my life. I've spent more than half a century in this place, in this undistinguished, comfortable town house on the edge of the Chiltern Hills, and had come to think we'd reached a pretty good accommodation. To have all mod cons on the doorstep of the quirkiest patch of countryside in south-east England had always seemed just the job for a rather solitary writing life. I'd use the house as a ground-base, and do my living in the woods, or in my head. I liked to persuade myself that the Chiltern landscape, with its folds and free-lines and constant sense of surprise, was what had shaped my prose, and maybe me too. But now I'm upping sticks and fleeing to the flatlands of East Anglia.

My past, or lack of it, had caught up with me. I'd been bogged down in the same place for too long, trapped by habits and memories. I was clotted with rootedness. And in the end I'd fallen ill and run out of words. My Irish grandfather, a day-worker who rarely stayed in one house long enough to pay the rent, knew what to do at times like this. In that word that catches all the shades of escape, from the young bird's flutter from the nest to the dodging of someone in trouble, he'd flit.

Yet hovering on the brink of this belated initiation, all I can do is think back again, to another wrenching journey. It had been a few summers before, when I was just beginning to slide into a state of melancholy and senselessness that were incomprehensible to me. I was due to go for a holiday in the Cevennes with some old friends, a few weeks in the limestone *causses* that had become something of a tradition, but could barely summon up enough spirit to leave home. Somehow I made it, and the Cevennes were, for that brief respite, as healing as ever, a time of sun and hedonism and companionship.

But towards the end of my stay something happened which lodged in my mind like a primal memory: a glimpse of another species' rite of passage. I'd travelled south to the Herault for a couple of days, and stayed overnight with my friends in a crooked stone house in Octon. In the morning we came across a fledgling swift beached in the attic. It had fallen out of the nest and lay with its crescent wings stretched out stiffly, unable to take off. Close to, its juvenile plumage wasn't the enigmatic black of those careering midsummer silhouettes, but a marbled mix of charcoal-grey and brown and powder-white. And we could see the price it paid for being so exquisitely adapted to a life that would be spent almost entirely in the air. Its prehensile claws, four facing to the front, were mounted on little more than feathered stumps, half-way down its body. We picked it up, carried it to the window and hurled it out. It was just six weeks old, and having its maiden flight and first experience of another species all in the same moment.

But whatever its emotions, they were overtaken by instinct and natural bravura. It went into a downward slide, winnowing furiously, skimmed so close to the road that we all gasped, and then flew up strongly towards the south-east. It would not touch down again until it came back to breed in two summers' time. How many miles is that? How many wing-beats? How much time off?

I tried to imagine the journey that lay ahead of it, the immense odyssey along a path never flown before, across chronic war-zones and banks of Mediterranean gunmen, through precipitous changes of weather and landscape. Its parents and siblings had almost certainly left already. It would be flying the 6,000 miles entirely on its own, on a course mapped out – or at least sketched out – deep in its central nervous system. Every one of its senses would be helping to guide it, checking its progress against genetic memories, generating who knows what astonishing experiences of consciousness. Maybe, like many seabirds, it would be picking up subtle changes in air-borne particles as it passed over seas and aromatic shrubland and the dusty thermals above African townships. It might be riding a magnetic trail detected by iron-rich cells in its forebrain. It would almost certainly be using, as navigation aids, landmarks whose shapes fitted templates in its genetic memories, and the sun too, and, on clear nights, the big constellations – which, half-way through its journey, would be replaced by a quite different set in the night sky of the southern hemisphere. Then, after three or four weeks, it would arrive in South Africa and earn its reward of nine months of unadulterated, aimless flying and playing. Come the following May, it and all the other first-year birds would come back to Europe and race recklessly about the sky just for the hell of it. That is what swifts do. It is their ancestral, unvarying destiny for the non-breeding months. But you would need to have a very sophisticated view of pleasure to believe they weren't also 'enjoying' themselves.

When that May came round I was blind to the swifts for the first time in my life. While they were *en fête* I was lying on my bed with my face away from the window, not really caring if I saw them again or not. In a strange and ironic turn-about, I had become the incomprehensible creature adrift in some insubstantial medium, out of kilter with the rest of creation. It didn't occur to me at the time, but maybe that is the way our whole species is moving.

*

So, about to become a first-time migrant myself, I can't get that fledgling swift out of my mind. This sudden swoop out of the nest and into the huge skies of East Anglia isn't something I've chosen or planned. Maybe some long-postponed maturation programme is guiding me, but it feels more like a cascade of dice-throws. To put it briefly, for now: I came to a kind of 'finish' in my work (but certainly not in the rest of my 'business'), drifted into a long and deep depression, couldn't work, used up most of my money, fell out with my sister – my house-mate – and had to sell the family home. Coming through was just as serendipitous. I was rescued by friends and slowly renovated, like an antique typewriter. I fell in love and started to write again, though with no idea of what I wanted to say. Then I caught a chance, as casually and as unexpectedly as one might a breeze. A couple of rooms in a friend's farmhouse happened to become vacant in East Anglia, which I'd seen as my second home since I was a teenager. Roofless and jobless, I jumped, and started again.

Now, packing the car, I feel like a *tabula rasa*, stripped down and open for offers. Even my belongings are, in both senses, spare. (I don't, for instance, have a single cooking utensil, telling myself they'll be 'provided' or at least available in my new habitat.) I have the tools of a trade whose survival value is debatable: a couple of manual typewriters and a drawerful of office gadgetry. But beyond that, my baggage is strictly sentimental. It includes a crystal of melon amethyst from Zambia, given to me for luck by my companion Poppy. A Victorian brass microscope, magnification approximately 100x. A picnic hamper full of elegant willow-pattern plates and cups, too posh ever to have been used. A badge inscribed with 'Cat Lovers Against The Bomb'. A sizeable chunk of the 1,500-year-old Selborne yew, which I have clung onto since it was blown down in 1990, convincing myself that I'm just waiting for the 'right carver'. Mum's favourite book, John Moore's *The Waters Under the Earth* (which if I'm right about the East Anglian landscape, may soon be mine too), with the Oxendale's catalogue order form she used as a bookmark. Emblems and fossils. I might just have well packed a pair of man-sized crescent wings for all the use these romantic knick-knacks will be. And as for books, I've sifted out a couple of hundred essential volumes (including most of John Clare) and sent the rest into storage in an industrial container somewhere up the Great North Road.

What a way to start a new life. I don't think I'm in denial, or 'downsizing'. This baggage, condensed into a few boxes in the back of a jeep, is actually all I want, and, to tell the truth, all there will be room for where I'm going. But I can't avoid the hugeness of change. This move is the thing I've been scared of all my life: the rite of cutting the cord, leaving the nest, spreading one's wings. It's a process so universal that we scarcely ever refer to it except in metaphors from nature. The only problem is that I've postponed it for a ridiculous and unnaturally long time.

Yet now the moment of severance has arrived, I'm feeling oddly elated and, for a dare on myself, I drive past the old house. The new owner's grandmother and her grandchildren are strolling round the garden, inspecting the remains of my old roses. It's odd staring at a scene which I've enacted so many times myself, both as a child and an adult, and knowing that I'll never make that ritual beating of the bounds again. Yet I don't feel in the least bit unreal, or as if I'm having an out-of-the-body experience, gazing in at my past like this. It seems instead almost a comforting image, of a kind of bequeathing.

It's a bright, balmy October day and feels more like the beginning of a summer holiday than a rite of passage. The fields, just free of a sharp frost, look burnished under the sun. Near Royston, a flock of lapwings, migrating south, veer over the road, and I remember the last time I saw them at a moment of change, a brief glimpse then, as now, of transience. I'd been up on Shap Fell with the photographer Tony Evans, searching for bird's-eye primroses, and the lapwings had flown – that loose, wavering flight, like windblown paper – over the honey-coloured pastures exactly above the point where we found the flowers. It was a sign of the year's turning and the last stages of a book we had worked on together for six years.

Only one thing sours the day. Somewhere back in my old home town there is a funeral pyre of the family furniture. It was of no value, just sensible, utility stuff our parents had bought for their first house. They were migrants themselves. Born and bred in London, they had premonitions of war and had headed west to make a home in the market town of Berkhamsted in the Chilterns. Sixty years later the pieces they had put together and with which we'd lived contentedly had become so much lumber. The house clearance firms refused them, and would not take them away even for cash. A charity furniture recycling centre ('sorry we don't take messages' said its messaging service) was dismissive. So most of it went on the fire, shouldered out of the house in a hurry by friends. The evening after it went was the only time my sister and I broke down. Between us we had 110 years of dwelling here, and sitting in the empty and echoing dining-room, with just one chair between us, was like being orphaned, or losing one of your senses. Because it wasn't the bare shell of the house that held the memories of the place, but its material things, the ordinary currency of living. The sideboard, whose top was already worn into hollows by two generations of fingers. The low, battered wicker chair on which Mum had nursed all four of her children. A cast-iron money-box-turned-doorstop in the form of an owl ('PAT.SEPT 21 & 28 1880') which for fifty years had tripped up everyone who came through the sitting-room door. 'Mind the owl' rang round the house, every time. Things become a kind of external memory, an embodiment of events and feelings.

My road east has that sort of engrained feeling too. I've been travelling it since I was a teenager and its swerves and shifts of character have become old familiars to me. When I first started to travel to the coast with a group of friends in the early 1960s, there was a precise point where we believed East Anglia began. Just out of Baldock, about 30 miles north-east of the Chilterns, we turned into the route of the neolithic road known as the Icknield Way. We passed a handsome Victorian maltings (now demolished). In front of us was a vista of sweeping chalky fields and a sky full of larks, and the modest prospects of the Home Counties seemed irrevocably behind us. It was a kind of portal, the point we knew we were finally 'away'. East Anglia spread before us like the awkward corners of a room that no one bothers to sweep, and we took to referring to it as 'the Ankle'.

Beyond Newmarket, the Icknield Way crosses the region known as Breckland, the dry heart of the region. Breckland is a 400-square-mile bowl of chalky sands. The thinness of the soil meant that the natural woodland cover was easy to clear, and for a while this was the most densely inhabited area of prehistoric Britain. The early farmers practised slash-and-burn agriculture, growing crops for a few seasons and then allowing the barely fertile land to 'go back' for twenty or so years. These briefly cultivated, 'broken' plots were known as brecks, and gave the region its name.

The sandy soils could barely sustain even this light cultivation, and when rabbits, and then sheep, moved in, the vegetation over much of Breckland became sparse. Up till the nineteenth century it was pretty much a dust-bowl, a wilderness of blown sand, and was so desolate that travellers used to cross this 'vast Arabian desert' at dawn, to avoid upsetting the horses. There was even an inland lighthouse to guide anyone who became benighted. The diarist John Evelyn wrote that 'the Travelling Sands have so damaged the country, rolling from place to place, and quite over-whelmed some gentlemen's estates' and he urged them to plant 'tufts of firr' to stabilise the sand. The gentlemen of East Anglia didn't need much encouragement, and during the eighteenth and nineteenth centuries Breckland was tamed, ploughed and enclosed. The only reminders of its old wild spirit are the occasional dust-devils that rise in the carrot fields and duck-ranches and blow across the lanes, and the lines of hunchbacked, windbreak pines that were planted and laid to keep the sand in its place. (This mutability is the subject of a classically stoical local joke. 'What county's your farm in?' 'That depends on which way the wind blows. Sometimes thas in Norfolk, sometimes in Suffolk.' This is East Anglia's creation myth: a world built on shifting sands.)

What's left of Breckland has, like so much so-called wasteland, become the dumping ground for the kind of land-uses that people don't want near their settlements. It's the site of nuclear strike airbases, the 25 square miles of the Ministry of Defence's Stanford Practical Training Area, and the earliest Forestry Commision conifer plantations – a new generation of 'tufts of firr'. Moving across this arid, sandy heart of East Anglia you are travelling through not just superficial changes in the scenery, but whole strata of shifting cultural attitudes towards the land.

Breckland's sense of barrenness, of isolation, has made it into a kind of frontierland, and what hasn't been written off as institutional wasteland, has been staked out, appropriated as a private playground. North of Bury the big estates are dedicated these days to horse-ranching or pheasant-rearing, and throughout the autumn and winter, you must drive willy-nilly over carpets of squashed, battery-reared birds that were quite unprepared for life in the wild.

Only the flints are constant, dogging your route and their chalk motherlode all the way to the north Norfolk coast. They're scattered uncountably in the fields, and built into the fabric of this whole swath of England: in cottages, walls, churches (round or flat side to the sun, depending on the district) and in multitudes of knapped artefacts found in the fields – arrowheads, axes, grinders, and the prototype, pear-shaped, all-purpose crunchers and cutters known as 'coups de poings'. The mine where the finest of them were dug, Grimes Graves near Thetford, is only 20 miles from where I'm headed.

We used to walk barefoot on flints for dares when we were kids, and this run through eastern England – over the sharp stones towards the sea – has always seemed a natural extension of childhood play. But today, instead of heading for the coast as usual, I take an unfamiliar route towards the region's heartland. South and east of Norwich are the rivers Waveney, Bure and Yare. For thousands of years, as the level of the North Sea rose and fell, they periodically flooded the region, and turned much of it into a swamp. A great fen stretched all the way from the brackish marshland of the three rivers' common estuary at Yarmouth, inland for maybe 40 miles, right up to the edge of the Breckland sands. A swamp of reed, lagoons and sagging alder woods, seething with life, with spoonbills, ospreys, cranes and otters, and creatures like pelicans and beavers that are now just memories. Even the human inhabitants had to be amphibious, and in the most waterlogged places they built their houses on raised platforms and walked about on swampland equivalents of snow-shoes called 'pawts'. In the east, during a spell when the water table was comparatively low, the locals dug extensive peat mines in the river valleys – which flooded to form the Broads when the sea-level rose again. To the west, where the three rivers rise, the fens and wet valleys were also mined, but were smaller and more isolated. One of these, the Upper Waveney Valley, is to be my future home.

*

Turning down that road less travelled, I can't any longer duck the questions which have been so unsettling me for the past few months – and in a more general form, I suppose, for much of my life. Where do I belong? What's my role? How, in social, emotional, ecological terms, do I find a way of *fitting*?

Back in the Chilterns I saw my home as a mooring, an emotional refuge, but used it as casually as if it were an old coat. Outside was where the real business was, either in the landscape itself or, metaphorically, in books. Living the free-range life of a writer, I skimmed about the countryside where and when I liked, following my own footsteps or taking off somewhere utterly unknown, hiding-up or socialising when the weather was bad, and avoiding heavily farmed countryside like a plague. I was a scrap of nomadic tissue, a kind of mobile epiphyte – an organism without its own roots – living on the land rather than in it, and letting others bother about my infrastructure. And like an epiphyte I was lost when my substrate started to collapse. I was, quite simply, too specialised, and didn't have the flexibility and confidence to cope with changes in my niche.

Up in the East Anglian borderlands I know I'm going to have to confront the daily realities of country life in a way I never have before. The weather, for a start. In Norfolk it comes famously straight from the Urals – and in an old house, I've no doubt, straight down the drive and in through every door, window and penetrable cranny. So, I guess, will big farming. The valley itself is a narrow tunnel of wildness. But I'm under no illusions about what lies beyond. East Anglia is the most intensively farmed region in Europe, and agriculture comes right up to three sides of the house. There is not what I would call a wood within 5 miles. Instead there will be rape and sugar-beet prairies, battery farms, condominiums of silos, *battues* of pheasants and pesticide drift.

And beyond finding a *modus vivendi* with farming I wonder how I'm going to cope with this bare and quintessentially watery place. I've spent most of my life under the cover of trees, and the rhythms of the woodland year have been a kind of metronome: that burst of brilliance and energy in the spring, the long, dense trance of summer, the sumptuous fading-away of the autumn, the clean, bare months of winter, the season of working and cutting back. I reckoned I could always lose myself in a wood, not just in its spaces, but in the layers of history bound up in its grain and forkings and slow cycles of light and shade. A wood's aura of history seems to go back not just beyond what generations of humans have done to it, but beyond civilisation altogether. The forest, the wildwood, is the nature we think we have, in both senses, 'grown out of', and in woods you have always the feel of 'going back'. They are places of long memory, and resilience.

What the sense of wildness will be like in my new surroundings is something I'll discover. But I've seen enough of wet places to know that they can be mercurial and unpredictable. By contrast with the cryptic, measured rhythms of woods, they have a vividness and immediacy, a sense that they might at any moment turn into something else. Very often they do. The wet is older than the wood, but it is the domain of the present, and sometimes, it feels, of the future.

The wood and the water, the ancient and the opportunist, are, I suspect, the two poles of natural rhythms. Life begins in the water and reaches its full maturity in the forest – and then it all goes round again. I think, greedily, that I'd like a bit of both, to become amphibious myself and to see that woodland high moment, spring, acted out in a swamp. But I wonder if I've got the stomach to live that much more intimately with any kind of wild. As unfamiliar creatures crowd into my new territory (and me into theirs) and as I try to adjust to hard-bitten ideas about utility and productivity, what will happen to my sense of the value and meaning of nature? Is hoping for a relationship that transcends the functional futile in such circumstances? In fact is the idea of a 'relationship' just playing with words, since nature seems not to have the slightest interest in having one with me? Should I settle instead for the time-honoured role of local naturalist, check out the birds at the garden feeders and try to add a few plant species to the county lists? Aspire to be a clerk of the records?

I said that this edgy baptism – which looks like being a real dunk in water – is about finding my place. Yet the whole business – fitting in, sharing territory, discovering a niche, making, with luck, a contribution, and trying to do it all with a modicum of grace and inventiveness looks uncannily like the challenge our species is facing as it tries to find its own settlement in nature. The difference is that, ecologically and globally, we're bucking the whole emotional aspect of that settlement. So often we're lectured that the great environmental crises of our time are just problems of household management writ large. If we're less greedy, stop breeding, budget our energy use, recycle our waste, make compost, then everything will be fine. What a hope! Who could ever run a house (if we must use that bossy, domestic metaphor) while ignoring the unquantifiable tastes and habits, the needs and motivations of all its occupants? The list of our disastrous failures, from forest obliteration and oceanic pollution to the raising of the extinction rate a thousandfold, bears all the marks of a species which no longer believes itself to be part of the animal world at all. We're becoming unearthly, freed, we like to think, from the physical imperatives of nature by technology, and exiled from its sensuality and immediacy by our self-awareness. Our role on the planet is compromised less by our power than by this arrogance, and the belief that our particular brand of consciousness makes us uniquely privileged as a species, entitled to evaluate and manage the lives of all the others on our own terms.

I've seen this high-mindedness close-to, and not been above it myself. I said earlier that I'd grown up under cover of woods. But it was more than that. For twenty years I had a wood of my own, to the extent that anyone can possess a wild community, a whole place. In that time I went through all the rites of property ownership, turning from inveterate trespasser to fence-keeper and fastidious surveyor, and ending up as dispossessed as all property owners eventually are. The upheaval my illness caused had brought my relationship with the wood to crisis point. I had no stomach to be an absentee landlord, and, to be frank, needed the capital tied up in it. I was having to sell up, and felt bitter and guilty and not at all resigned to it.

     I'd bought Hardings Wood, a 16-acre patch of ancient forest 800 feet up, near the village of Wigginton in the Chilterns, back in the early 1980s, and for two decades it had been the apple of my eye. I'd wanted to set up a community wood project and free-up what had been a dark and private timber-lot for the benefit of the local community, of all species. It had been, in those terms, a spectacular success. Half the village had worked or walked in it.

Extravagant sheets of woodland flowers and self-sown trees flourished in the light, and a dynasty of badgers spread through a network of catacombs below. A new consciousness of the wood as a landmark, a place of both dense history and great presence, began to take root in the area. But now, numb at the thought of losing it, I wasn't admitting a lot of other, more personal business. For me it had also been a playground and a stage. I'd written about its bluebells and the Ascension Day celebrations held there when they were in full flower. I'd been filmed wielding a chain-saw with criminal abandon. Above all, I'd used it as an immense private library of experiences and encounters.

It had been a lesson in social relations, too. I learned, intimately for the first time, the extent to which the ethos of property permeates our relationships with the natural world. I took down one lot of fences to let the public in but had to put up another lot to keep my neighbours' animals out. I crossed my fingers and signed up to 'control' grey squirrels in order to get a government grant. I banned the hunt after it assumed it had the right to ride through the wood and kill foxes that I knew as individuals. Every time someone appropriated the place or its inhabitants as their property, I retorted in some feeble reflex, that they were *mine*.

Once I had my own enthusiastic largesse thrown back at me. A new road was scheduled to come through the wooded valleys where I'd played as a child, and pass within 550 yards of Hardings. I gave evidence at the public inquiry and mentioned, among other things, the colonies of bankside wood anemones that might be lost to the road. The counsel for the highways authority pounced. Hadn't I written about the anemones in my wood, put on record how abundant and luxuriant they were? How could the roadside flowers be of any value at all when the species was, by my own account, so common nearby? The lawyer wasn't being particularly devious. He was using the same argument about value employed as a matter of course by conservationists: plants aren't important in themselves, but only as representatives of a species. Their significance increases only as they become endangered. To care about individuals and the complex relations they have with their intimate surroundings and local ecosystems (the only relationships that matter to the wild things themselves, after all), is to be accused of being subjective or, worse, sentimental.

It's here that I have to jump ship. I've never been able to fit my feelings about nature into this kind of value system, weighing its usefulness and scarcity value like some kind of commodity dealer. I like common things, and the idea of commonality – 'a Council of all Beings'. I find it hard to think of nature as purely a human resource, even though I know I physically depend on it, and harder still to perceive the non-human world objectively, as an 'object', when I know it has its own subjective tasks and goals, independent of yet permeating ours. And why should we try to be neutrals, when we are so inextricably and passionately involved?

And worse, I *am* sentimental. I talk to birds. I mark much of my sense of time and place with odd moments and fragments of the non-human world. I have a drink for the first swallow, keep a tape somewhere of a nightingale I recorded in a thick fog to play down the phone to a far-away girlfriend. I'm touched by the durability of these seasonal encounters, yet also by those moments when nature is breaking the rules, freeing itself from our tidy categories and timetables, being independent, unpredictable, sassy, making things anew. Both kinds of experience seem to me varieties of 'wildness', and far removed from the predictability of a man-managed world. One, of the deep, encoded, learned-experience of evolution; the other, of newness, invention, individuality, of the renewal of spring and the indulgence of play – yet of random pain and catastrophe, too.

*

So I'm thinking again about my feelings for that beached, fledgling swift. Where did they come from? I don't eat swifts, or aspire to have them as pets. I don't feel they need my protection, having existed perfectly independently on the planet for millions of years. And in a view of the world based on 'resource conservation', swifts are almost certainly irrelevant. They are not (yet) endangered. No important predator depends on them (and if it did, we would then have to ask what use that is). The planet's linked ecosystems – what James Lovelock has christened Gaia – would give no more than a sigh if they vanished. It would be testing credulity to suggest that one day they might be the source of a drug against, say, air-sickness, distilled from their prodigious balancing organs; or that their scrabbled-together nests (made from flotsam gathered from the air) could provide the inspiration for low-cost building designs. No, swifts do not pass the critical tests of endangerment or usefulness.

Yet they touch and connect with us in deep and subtle ways. We do not know what it would be like to live through a summer without them. They are part of our myths of spring and the South, a crucial element in that great gift to the temperate zone, the migration and settlement of summer birds. 'An annual barter of food for light', Aldo Leopold wrote of migration in America, in which 'the whole continent receives as net profit a wild poem dropped from the murky skies.' They are the most pure expression of flight, an ability which is still remembered somewhere deep in our nervous system. Swifts have become, I think, our twenty-first-century equivalent of the Romantics' nightingales – cryptic, rhapsodic, electrifying – but happy to be all these things at high speed in the middle of an urban landscape. Like the nightingale's 'darkling' song, the swift's black silhouette and utterly aerial existence give them a pliability of meaning.

When I was at school I longed so much for their return that I used to walk about on May Day clutching my blazer collar for luck. Later, when I was about seventeen, they became romantic emblems of high summer. I sang in an early music choir in those days, and on June evenings we rehearsed in the local parish church with the local girls' school perched on the opposite side of the chancel. The swifts screamed round the tower, and past the stained-glass windows lit by the low sun – a shrill descant to our own warblings. It was a haunting scene of unrequited courtly lust, and though swifts are now beyond the range of my hearing, I can still recall the sound of those evenings, along with the forbidden thrill of the girls' green gingham dresses.

In my adult life swifts have become more mysterious, not a symbol of anything in particular, but creatures made of the same cells and tissues as me yet living on another, almost unknowable plane. Their existence in – and sometimes, it seems, on – air is more cryptic than the livelihood which myriads of creatures make suspended in and supported by water. Swifts eat, sleep and mate on the wing. They gather windblown debris for nests and bathe in the rain ('they take showers', wrote William Fiennes). All over Europe – in that extraordinary town of birds, Trujillo, and lost on a motorway somewhere just outside Montpellier – I've stood mesmerised as they hurtled past me at waist level, wondering what they made of me. Were they aware that – earthbound and ponderous – I was even alive?

Mostly I used to watch swifts down by the canal in my old home in the Chilterns. I'd go to a pub an hour or two before sunset, and lose myself in their vespers ritual. They nested, a dozen or so pairs, in the eaves of a row of Victorian terraced houses, and in a disused factory that used to manufacture insecticides. And on warm, still evenings the young neighbourhood singletons would form a loose flock, trawling insects a couple of hundred feet above the town centre. They looked as if they were milling about haphazardly, like ash specks over a bonfire, but would cross and swoop across each others' paths without missing a wing-beat. Then some ancient compulsion took over, quite pointless unless you are prepared to grant birds a sense of pure physical relish in their power of flight. At the edges of the swarm the birds began to wheel, wings stiff. One by one they dived down to a lower altitude, and began to fly and chase, first in pairs then in accumulating strands, until maybe thirty birds were careering together in a mass. They became a ragged black comet, fizzing with activity as birds throttled back, feathered, did spectacular left-right wing-tilts to avoid collisions. The comet hurtled between the factory buildings, calling to the sitting birds inside, banked like a gang of motor-cyclists to take the curve back over the new wharfside flats. They seemed to be following tracks in the air that only they could see – but would then spectacularly run off them. They'd disappear only to somehow reappear behind me. Then abruptly, at no visible signal, they would fling apart and fly off leisurely in different directions, making for the high air again.

I never saw the actual moment when they disappeared for rest. I suspect they gained altitude rather like an aircraft, flying out of the town on a gradual gradient. But I have seen a film of the image south-east England's sleep-bound swifts make on air-traffic-control radar. As darkness falls all the aircraft on the screen are obliterated by an ethereal halo of bright coalescing spots, each one an individual group of swifts, bound for that state of total otherness – invisible, aerial slumber.

As a relationship, my thing with swifts is so one-sided as to be hardly worthy of the name. The birds don't give a fig about me or any of us. Yet they are connected with us indirectly, even when we are not aware of them, through the environments and senses that we share. We respond to the spring, to the lift of fine weather, to the basic biological urge to play. In his poem 'Swifts', Ted Hughes writes of his feelings on their return ('They've made it again'), and how they symbolise not just the summer's renewal, but that 'the globe's still working'. It's a common response. I was once sent, out of the blue, this short perfect poem, scribbled on a sheet from a memo pad by Margaret Thomson:

ASCENSION DAY

May. Just into
Double figures.
Everything green
And brilliant –
The first warm day.
Soft shoes, no socks,
Then you calling out
'The swifts are back!
Listen. Look up!'

Listen. Look up! Did birds like swifts, arriving mysteriously in the spring, reappearing from nowhere at dawn, play their part in the generation of resurrection myths? Do they still register, at the corners of our vision and reason, something immanent? Despite our science and our humanism, our whole culture is infused with myths and symbols of landscape and nature, emblems of the seasons, of decay and rebirth, of the boundaries between the wild and the tame, myths of migration and transmigration, of invisible monsters and lands of lost content.

We constantly refer back to the natural world to try and discover who we are. Nature is the most potent source of metaphors to describe and explain our behaviour and feelings. It is the root and branch of much of our language. We sing like birds, blossom like flowers, stand like oaks. Or then again we eat like gluttons, breed like rabbits and generally behave like animals. But then 'animal' itself springs from the ancient Sanskrit root *anila*, meaning 'wind', via the Latin *animalis*, 'anything alive', splitting off *animus* on the way as, first, 'mind' and then 'mental impulse, disposition, passion' – a reminder of the time that mind and nature were not thought of as contrary entities. It is as if in using the facility of language, the thing we believe most separates us from nature, we are constantly pulled back to its, and our, origins. In that sense all natural metaphors are miniature creation myths, allusions to how things came to be, and a confirmation of the unity of life.

Edward O. Wilson popularised the word 'biophilia' for this pervasive affinity our species seems to feel for others. In his definition it is 'the innate tendency to focus on life and life-like processes'. The wild creature most anciently focused on in East Anglia is the hare, the mysterious, prankish 'stag of the stubble, the looker to the side', which has been, variously, a witch's familiar, an emblem of spring, a fertility symbol, a moon-creature, a fire-devil and a trickster. Maybe its shape-shifting antics in the field are what have made it one of the oldest and most widespread animals in world mythology. Up here it used to be thought bad luck if a hare crossed the road ahead of you. But the poet Cowper's journal of his hare Puss in 1786 is one of the great accounts of the mellowing relationship between humankind and animals. Brer Rabbit was a hare. Variations on the parable of the tortoise and the hare can be found in almost every language from Bantu to Tibetan. Across parts of North America the Great Hare was the lynch-pin of creation myths. And an Egyptian hieroglyph from 2000 BC of a hare over a ripple of water meant simply 'to exist'. There is a touching Chinese story, a very ecological folk-tale, that weaves together many of these symbolic roles. It concerns a hare, living in the

Buddha's sacred grove, whose virtues gave him ascendency over all other animals. One evening the Buddha came disguised as a starving Brahmin. The hare bounds to his aid. 'Master, I who have grown up in the forest, nourished by grass and herbs, have nothing to offer thee but my own body. Vouchsafe me the favour of feeding thee with my own flesh.' He then throws himself on a charcoal fire, but not before stopping to gently pick off the fleas from his fur. 'My body I may sacrifice to the Holy One, but your lives I have no right to take.' As a reward the Buddha commanded that the hare's image should for ever more adorn the face of the moon.

But the imagining and mythologising of nature is an ambivalent process. Its own 'truths' can stand in the way of solid scientific 'facts'. Its metaphors and images and symbols can be a substitute for an engagement with the real world (sometimes permanently, as in 'dead as a dodo'). The hare, after all, is still hunted for 'sport', and its habitat still trashed by modern agriculture. More fundamentally the very facility of language that gives us the ability to think in this way is often seen as an impassable barrier between us and nature, the thing that estranges us, prevents our ever joining hands.

But I am not so sure. Once in my wood I had a face-to-face meeting with a female muntjac deer. It wasn't a sudden enounter, a collision round a bush and a moment of mutual fluster. We'd sidled up to each other, both with that slight tilt of the head that universally signals curiosity, caution, an uncertainty about what may happen next, and an unwillingness to be either provocative or provoked. We got to about 10 feet from each other and then just stared. I looked into her large eyes and at her humped back and down-pointed tail, which signified she wasn't alarmed. I thought about what race-memory she might have of her species' Chinese homeland, about what she made of an English beechwood and whether she'd looked at a human's face before. She looked at my eyes and passed her tongue repeatedly over her face, wondering if I was dangerous or musty. I thought she was pretty and plucky. She thought I was odorous and interesting and curiously shaped and things I could never know. We conversed in this way, two curious and diffident strangers, then went our separate ways. I returned home, and wrote up my memories of our meeting. She wandered off, probably consciously 'forgetting' it, but with intricate impressions of my scent and shape and the sounds of my breathing lodged somewhere in her mind.

The writer Ian Sinclair, from inspiration or mishearing, calls this animal the monkjack, a name which perfectly captures its solemn solitariness and bouts of mischief, and which helps when I try to make sense of our encounter. I can't make value judgements about what was going on, beyond saying that she had used her monkjackish skills, and I had used my human. But the encounter seemed every bit as 'natural' to me as if I had been hunting her or sharing her food.

I would like to get to know this valley as well as I knew my wood, to be as easy with, as *conscious* of, its dwellers as I was with that monkjack. This book will be a record of how things turn out during that first year. But it will also, inevitably, be an account of my own life in the aftermath of illness, and of what I felt and thought dipping my toe at last into something approaching adult independence. It's become customary, on this side of the Atlantic, stiffly to exclude all such personal narratives from writings about the natural world, as if the experience of nature were something separate from real life, a diversion, a hobby; or perhaps only to be evaluated through the dispassionate and separating prism of science. It has never felt like that to me, and since my recovery, it's seemed absurd that, with our new understanding of the kindredness of life, so-called 'nature writing' should divorce itself from other kinds of literature, and from the rest of human existence.

Learning to write again was what finally made me better – a story that has its proper place later – and I believe that language and imagination, far from alienating us from nature, are our most powerful and natural tools for re-engaging with it. I hope that what happens to me in my new habitat will reinforce that belief. Culture isn't the opposite or contrary of nature. It's the interface between us and the non-human world, our species' semi-permeable membrane.

*

I'm half a mile away now, and the pink farmhouse is slipping in and out of view between the shrubs on the common. A few gorse bushes are still in flower, and the fading heather blooms are giving a spent, rusty tinge to the ground. Beyond, a thickening of willow and alder marks the edge of the fen and the river. I've been along this road just once before, yet everything about it is familiar. I grew up on commonland, among the same mixture of furze and birch, and I recognise the patterning of the paths, separating, funnelling, joining again. They're the marks of solitary animals with incurable habits, of moments of sociability, of the crossings of paths of different creatures. They're the signature of every stretch of commonland on the planet.

I know it second-hand, too, as the kind of habitat in which the nineteenth-century poet John Clare lived, worked and wrote. Clare was one of the few writers to have found that shared field, and to have created a language that joined rather than separated nature and culture. (He once said that he 'found my poems in the fields'.) He felt most at home in the open, tangled, undramatic spaces of wasteland and heath. He saw the whole living landscape as a kind of common and himself as just one of the commoners.

'Unbounded freedom ruled the wandering scene', he wrote in his elegy for the open places in his native village lost to Enclosure:

> Moors, loosing from the sight, far, smooth, and blea,
> Where swopt the plover in its pleasure free
> Are vanished now with commons wild and gay
> As poet's visions of life's early day

I feel a bond with Clare. He is – almost – an honorary East Anglian, and had his own struggle with depression. For a mercifully brief spell, I had the eerie experience of being in the same hospital as he had been in 150 years before. He stayed there. I got out. But he is still a companion, an urgent, troubled part of me, dodging from bush to bush, and I need to follow his trail, too.

2

Lair

LAIR: 1.c A place for animals to lie down in.
4. Of land. The state of lying fallow.
Oxford English Dictionary

I'D BEEN GOING TO earth most of my life. When I was about six, I dug holes in the garden – big holes, the size of a treasure chest – and curled up in them. A few years later it was the inside of trees. If I ignored the scratching and burrowed in deep enough, they all seemed to have spaces inside where the mass of twigs had left a calm hollow. Once our neighbourhood gang made a den in the root-hole of a wind-thrown sweet chestnut. It was as big as a cave inside, scented with the spicy aromas of damp clay and torn roots, and one day we tempted our friend Ann in and tried to get her pregnant by poking nervously at her chest with a sprig of holly. Or so I was told. I was deemed too young by the tribal elders to indulge in such serious pagan business.

But the arrival of real sex made such hideaways essential. I had favourite retreats for my own private experiments in a marlpit in Bulbeggar's Wood on the lane from Berkhamsted to Potten End and, more brazenly, up against various voluptuous trees on the way to cricket. And soon, there were thrillingly dangerous encounters with my first girlfriend ('the jewel of Berkhamsted' her mother called her, as she stood guard over us) in a dark and musty pillbox about 3 yards from a road through our town centre. When I lived in London for a year after university, I couldn't bring myself to get something as respectable and established as a room of my own, and squatted for three months in a friend's broom cupboard. Even in my thirties, when I had a whole wood of my own to play in, I still nosed out sanctums inside holly and bluebell glades where I used to go into retreat. Into retreat from what I am not sure, except that the feeling of being hidden, unknown, maybe untraceable, was exhilarating.

Did these lay-ups, as different on the surface as caves and cloisters, have anything in common? I think they were all places which briefly combined security and adventure; which were retreats to look out from as much as be in, breathing spaces for gathering myself or rehearsing — all those pairs of qualities that landscape theorists refer to collectively as 'prospect and refuge'. Yet the one place that was not a lair, that was all refuge and no prospect, was my old home in the Chilterns. That had been a womb, a hermitage, a place I had used to keep the outside world at bay, and growing out from it was my chief unfinished business.

But I took comfort in my lairing habits from the behaviour of other creatures. The sculptor David Nash has painted a series of what he calls 'sheep spaces', those nooks under tree-roots or in the lee of rocks that all animals scrape out when they need rest or shelter. Cock wrens do something similar in early spring, making half a dozen exquisite play nests before the big performance that is meant to lure in their mate. In and out go the meticulously woven moss and feathers, round the little entrance John Clare likened to 'a cork-hole in a barrel'. But it is all so much bachelor stuff, done for practice, amusement and showing off.

*

It was a lair, more or less, that I'd fetched up in, a happenstance, temporary retreat. I was in no state for a proper roots-and-branches home yet. What I needed was something handy, make-do, maybe a touch cloistered. The terms I agreed with my landlady Kate were in exactly this spirit. No permanence. No serious baggage, either physical or emotional. We could barely bring ourselves to sign a contract, it smacked so much of formal colonisation. This is what it roughly involved. While she worked in London during the week I would take care of her three cats, and a hutchful of hapless small mammals that I instantly consigned to an impenetrable wood at the back of my mind. I'd also keep a general eye on the house, receive parcels, admit plumbers, that sort of thing. And I'd attend, when I could, to the array of wild creatures that dwelt here: the hornets in the house and in the fen; the long-eared bats and swifts that occasionally bred in the loft and the thronging flocks of birds that came to the feeding stations.

I barely took in the setting of the place to start with, its remoteness, its insidious wetness, the sweeps of ling and gorse that had given the farmhouse its name. And for all I noticed of it in the first days, the house itself could have been perfectly summed up in estate agents' jargon: '17th C fmhse, lovingly rest, hlf-timbd, orig winds and flrbrds, 9 rms, suit writer/recluse'. I was in a state of culture shock, and had thoughts only for my first-floor room, and how I was going to re-learn the basic living habits of forty years in about two days flat.

Going into my eyrie was like walking into a small forest. It was built around and between a grove of petrified oak timbers, stripped and faded to the colour of old bone. It had 6-inch-wide oak floorboards and an oak desk. There was more oak inside the room than out. Outside the mullioned (oak) north-facing window, the meadow sloped down to a willow and alder carr beside the River Waveney. To the south was a view over a field of sugar-beet, rising gently up towards the Manor House woods on the ridge. Between them and my window was a small walled garden, with a pear tree and a couple of Irish yews. It was alive with birds, right up to my window pane. Pheasants stalked the lawn. Great spotted woodpeckers made beelines for the nut feeders, looping across the lawn or shinning up the pear tree. A pair of bullfinches (that flash of white rump and soft whistling) materialised in a rose bush. I hadn't seen a bullfinch for about six years, and standing in this awesome, *demanding* room in which I was supposed to make some kind of habitation, I thought of Ruskin's reassuring words about their nests. He'd been shown one made entirely out of clematis stalks, with the withered flowerheads all on the outside, like perhaps 'an intricate Gothic boss of extreme grace and quaintness, apparently arranged . . . with the definite purpose of obtaining ornamental form'. Alas no, Ruskin concludes. While the bullfinch isn't 'a mechanical arrangement of nervous fibre . . . impelled by galvanic stimulus to the collection of clematis', it isn't an architect either. It 'has exactly the degree of emotion, the extent of science, and the command of art which are necessary for its happiness'.

Those sounded pretty good prerequisites for life, let alone lair-building, and I tried to feel what the woody intricacy of the place (in which the oak's 'flowering heads', so to speak, also seemed to be on the 'outside', exposed and grain-visible) meant in terms of homeliness. What kind of nest was this, and for what kind of creature? What stood out most, as direct to the eye as an overhanging branch, were the six diagonal beams across the corners of the walls. Each of them had an identical bend of about 10 degrees, half-way along, and I was struck by the thought that they might all have been sawn from the centre of the same crankled oak. And that the tree had been chosen and cut with its future role in mind, maybe from woods nearby. I could see on at least three of the diagonals a smudgy, oval area of knot one foot down from the elbow. It had the look of a contour map, or the inside of an oyster. As the beams had been cut with the grain, the growth pattern wasn't visible as rings, but as a swirling pattern of longitudinal stripes and eddies. But near the edge they were bunched close. Dry weather grain. I thought of the room as a weather fossil, of maybe a spell of drought in the valley woods, four centuries ago, at the very start of the Age of Reason. I was happy to roost in a place from this undecided period, when imagination still had a role in the understanding of the world.

But I winced as I began to arrange my books and pictures and yards of flex – all straight lines and jazzy colours – against this ancient austerity. Why such titivation? it seemed to ask. Why not a bed, a chair and a bare table like the first owner would have had? I tried to imagine the modern equivalent, something Japanese and minimal, but had only one real answer, the same as he or she would have given. I had to live here, and make a livelihood. I needed books, I needed warmth and I needed light – or rather lights. The ceiling was low and timbered itself, and a single centre source would have been impractical. So I tried instead to build up some sort of ecosystem of individual low-energy lamps. I strung them round the room in the places where I would do things: an intense light for typing at night; a softer, dimmable one for resting and reading in the armchair; a mobile one for finding books, and for the dark corner where the radio and fax machine lurked. When they were all on at once it was unnervingly like a Christmas grotto, but, rationalising furiously, I preferred to imagine it as resembling a forest light-system, with shifting patterns of dapple and shine. Edward O. Wilson has described one such in Amazonia:

> The sun came out again and shattered the vegetative surfaces into light-based niches. They included intensely lighted leaf tops and the tops of miniature canyons cutting vertically through tree bark to create depths two or three centimetres below. The light filtered down from above as it

does in the sea, giving out permanently in the lowermost recesses of buttressed tree trunks . . . As the light's intensity rose and fell with the transit of the sun, silverfish, beetles, spiders, bark lice, and other creatures were summoned from their sanctuaries and retreated back in alternation.

I imagined making my own transit through the room's illuminated niches, foraging briefly in the CD player recess, retreating back to the glowing sanctuary of the chair.

So in the evenings I sat in a glade, frozen at the moment just before sunset. And one night, with the room lit with an orange haze, all the hard edges between flex and floorboards, laths and pictures, began to soften, and Elizabeth Blackadder's life-size print of the orchid *Paphiopedilum sanderianum* was suddenly growing *out* of the beam on which I'd fortuitously hung it. There it sat with the rootstock and leaf blades grafted smartly onto the wood, and the flowers' tail-streamers reaching down almost to the top of my bookcase. It had become a pictorial epiphyte, a mock-up of the real partnership that exists between orchid and tree. I felt forgiven, and thought of our house hornets, who were also recycling old wood into exotic décor. I hoped we were up to the same sort of business.

*

And there were the cats to think of too. They were in a disorientated and depressed state after four months in a cattery. Blanco, a sinuous white male with chrysoberyl eyes, and Lily, a middle-aged female torty, skulked, tense and immobile, under the very lowest of the beds in the spare room. Only Blackie, female, black and white, and I suspect a bit of an Essex cat, emerged, and with her disarmingly upturned face, was prepared to converse with me. I tried every dodge to coax the others out, but they remained resolutely determined to show that they were institutionalised and cross.

Along with the advice given in all respectable cat manuals, they were to be kept shut in their room for at least a month, and only then introduced to the labyrinthine spaces of the house and the countryside beyond. Fat chance. One morning, thanks to a combination of wind, ill-fitting doors, and builders' comings and goings, Blackie and Blanco escaped. They were nowhere to be found either in the house or in its immediate surrounds. I searched, getting increasingly frantic, in barns and sheds, down in the edges of the swamp, even, my heart sinking, along the lane. I was, I'm ashamed to say, as concerned about my own fate as much as the cats'; imagining my tenantship indelibly blotted and my role as Cat Reeve doomed. That evening, needless to say, they were back curled up in their room, with a dead shrew laid out ceremoniously on the carpet. Why on earth had I worried? Cats are less than four thousand years of domestication away from their wild ancestors, and have barely lost a whit of their instincts for independence and territorial adventure. After that first expedition they escaped into the house repeatedly, made forays half-way up chimneys, found my room and got locked in my wardrobe. We surrendered and gave them their freedom a fortnight in advance. They had – as Kate once offered me, pointing to the stairs when I was over-dramatising my change of life – '*rights* of passage'.

Following the cats' lead, I began to range around the recesses of the house myself. At night, a quarter of a mile from the nearest neighbour, I could behave as ferally as I wanted. I turned the lights down and the music up. I tried crawling on all fours down the narrow passageways, seeing if I could navigate by touch alone. In the small hours the place seemed only marginally removed from the woodland out of which so much of its fabric had been hewn. It was unquestionably still alive. Nothing was straight or on the level. As the temperature fell, the beams and joints contracted and the whole place shape-shifted. Ornaments, cups and phones slid to the edges of tables. Doors opened and shut mysteriously. If there was moonlight, the mullions cast stark branching shadows on the floors. One evening I went up to the loft with a torch. This was where the servants used to live, a ramshackle, wooden ridge-tent of ancient sheep hurdles and fallen thatch. In the north-east corner was the cobwebby corner where the swifts made their grass-and-feather nest. In two or three months, my totem birds might be snuggling down just 10 inches above the top of my wardrobe.

I was taken aback at my rapid plunge into house-consciousness, and at how domestic I was becoming. I was preoccupied with making the lair work. I began dyeing old clothes, improvising gadgets, doing risk-taking experiments in the kitchen with Hunza apricots and millet flour. I'd also turned fastidious about my things. I tidied them obsessively, before I started work, after I'd finished, and sometimes as I went along. I told myself that this was purely practical, that it was impossible to live and work in the same space without some order and system. But I knew that my bits and pieces, the trade tools and books especially, had much of my past and my present security tied up in them, and were very precious. I was behaving like any animal bedding down for the winter, checking essential stores, freshening things up, biding time till the new season.

I was feeling smug too about the efficiency of my writing niche, a lair within a lair, which I'd bodged and kitted up as snugly as the cockpit of a 1920s biplane. I'd fallen out with computers when I was ill, not from any ideological hostility but because I'd found the cornucopia of choices they offered at every turn too much to cope with. So instead I had, from port to starboard on two small desks, a pair of manual typewriters (one electric for smart copies), a fax machine, various jugglable phones, card-index boxes, that intense and unfluctuating spotlight, a pair of binoculars and, as an insurance policy, a very accurate maximum and minimum thermometer. I wanted to know when I was suffering.

People ask what it takes to be a full-time writer, to make it your livelihood, and the only honest answer is the doggedness to be alone in a room for a very long time, climbing without a rope. But working the edges of the niche, the way an incarcerated cat might, makes it more bearable. I found all kinds of useful diversions. I cleaned one of my typewriters using the nearest thing I had to surgical spirit – a bottle of Body Shop bergamot eau de cologne. I had the sweetest-smelling typewriter in East Anglia. Quite often, if I was bored or feeling particularly fleet of finger, I'd switch to the electric typewriter and be carried away on free associations from its seemingly spontaneous torrent of characters. My small expeditions to fetch books, due north over the woody ridge in the centre of my floor, were major translocations, as the south-facing vista of walled garden and farmland was replaced by the wilder prospect of meadow and fen to the north. When Annie Dillard was writing her extraordinary odyssey into the meaning of evolution, *Pilgrim at Tinker Creek*, she worked in a second-floor study carrel in Hollins College Library, overlooking a tar-and-gravel roof. 'One wants a room with no view,' she wrote, 'so imagination can meet memory in the dark.' She shut the blinds and pasted on them a drawing she'd made of the view from the window: 'If I had possessed the skill, I would have painted, directly on the slats of the lowered blind, a *trompe l'oeil* mural view of all that the blinds hid. Instead, I wrote it.'

*

In the evenings I have a room with no view, too. I sit in the chair with a cat or two, and go freeform reading. I trail ideas and names through my small hoard of books, giving in to the slightest opportunity to be side-tracked. I've followed the history of the marshland will-o'-the-wisps, poked my nose into the raffish lifestyles of Norfolk's writing dynasties, found one good new theory about animal consciousness, and disentangled the discovery and disappearance of the spectacular colour varieties of the local *Dactylorhiza* orchids. The jumble of marsh and wood and water in the fens, combined with the protean inventiveness of the orchid family, once nourished some true local prodigies. In the little swamp at Roydon, just a short distance from the house, J. E. Lousley found the first British specimen of the ethereal straw-coloured variety (*ochroleuca*) of the early marsh orchid. Now it, too, starved of water, lives only in books.

The dark seeps in (there are no curtains over the old windows), and I think about the oddness of a so-called nature writer spending so much of his time with his nose in a text. Shouldn't I be out in the night, sniffing for foxes, or peering for the dark shapes of deer in the fields, not busy with all this unnatural reflection and study? Isn't a life of words the very antithesis of a life of nature? Once, whenever I felt defensive about my social role, I'd argue that nature-writing was an authentic and honourable variety of rural work, dependent, like farming, on the weather and seasons and a fair bit of luck, and producing a crop of so many words per acre tramped or meditated upon. I'd flourish as evidence the clause in the tax return forms where the Inland Revenue grants a concession allowing the averaging of earnings over good and lean years – solely to 'farmers and creators of literary or artistic work'. What a coupling! But it was not an argument that went down well. Most people understand a writer's end-products, the odd concoctions of personal and social memories, the old myths and new metaphors, the smatterings of science. They will even accept you might have a useful role as a kind of local storyteller, one of the oldest jobs on earth. But the actual act of making the words remains an impenetrable and almost masonic mystery. 'But what do you do for a *living*?' non-writers are apt to ask, as if all these ethereal meanderings could never be regarded as anything other than play.

I often share their bewilderment. In one sense, writing and, for that matter, all imagining is play, an entirely gratuitous ornamentation of the real business of survival. It's easy to see it as just about the most *un*natural activity we do. But up in this wooded eyrie, surrounded by what looks like being a pretty feral landscape, I have a feeling that the whole thing is about to become more primal, more connected, more on a par with follow-your-nose hunter-gathering than orderly husbandry. I've found a book called, enticingly, *Enjoying Moths*. In it Roy Leverton describes how Victorian entomologists invented the word 'dusking' to describe their evening expeditions. They'd go out in the twilight with lanterns and sugar snares, hoping to lure the ghostly creatures into their nets. Hapless moths apart, I rather like the idea of dusking. It seems to fit my own evening sojourns, fluttering after a collection of elusive ideas that people have been trying to pin down for centuries.

The idea that spells of meditation and imaginings and writing-down aren't entirely divorced from the life of the wild has had some honourable advocates. Henry Thoreau was typically histrionic when he wrote:

> he would be a poet who could impress the winds and
> streams into his service, to speak for him; who nailed words
> to their primitive senses, as farmers drive down stakes in
> the spring ... who derived his words as often as he used

them, – transplanted them to his page with earth adhering to their roots; whose words were so true and fresh and natural that they would appear to expand like buds at the approach of spring . . .

But the great American poet Gary Snyder's elaboration of a similar idea is truly rooted. He argues in *Unnatural Writing* that

> consciousness, mind *and* language are fundamentally wild. 'Wild' as in wild ecosystems – richly interconnected, interdependent and incredibly complex. Diverse, ancient and full of information . . . Narratives are one sort of trace that we leave in the world. All our literatures are leavings, of the same order as the myths of wilderness people who leave behind only stories and a few stone tools. Other orders of being have their own literature. Narrative in the deer world is a track of scents that is passed on from deer to deer, with an art of interpretation which is instinctive. A literature of bloodstains, a bit of piss, a whiff of estrus, a hit of rut, a scrape on a sapling, and long gone.

Yet the issue is not so much whether or not human language is part and parcel of the larger creative business of nature (which of course it is, having evolved from the wild), but whether it resonates, chimes sympathetically with it. The idea that language is no more than our species' 'contact call', like the cheeps of winter bird flocks keeping in touch in the cold, is touching, but is really a relic of an older yearning for a return to innocence, and in a backhanded way reinforces the opposition of nature and culture. It is odd that the gifts of image-making and language are so often seen as the attributes which irrevocably alienate us from nature, are the cause of our Fall from grace. We will never know the state of self-consciousness of another species, but it's a reasonable bet that most don't use language in the way that we do, or think metaphorically, or mediate their vivid sense-experiences through such a complex net of associations and references.

Yet why on earth should this 'estrange' us? The notion that going 'back to nature' – longed for or dreaded, depending on your point of view – must necessarily involve some equivalent retreat from self-awareness is absurd. We have evolved as talkers and dreamers. That is our niche in the world, something we can't undo. But can't we see those very skills as our way back, rather than the cause of our exile? Of course, language and imagination have to some extent deadened the quickness of our sensual relationships with the outside world (though that is far from inevitable), and have made us aware of the ways in which we are a different kind of species. But they're also the gateway to understanding our kindredness to the rest of creation, to fitting our oddness into the scheme of things, to become awakeners, celebrators, to add our particular 'singing' to that of the rest of the natural world.

Jonathan Bate has written in *Song of the Earth*: 'The dream of deep ecology will never be realised on earth, but our survival as a species may be dependent on our capacity to dream it in the work of our imagination.' I'd go further and say that our imaginative affinities with the natural world are a crucial *ecological* bond, as essential to us as our material needs for air and water and photosynthesising plants.

In his autobiographical writings John Clare often writes of 'dropping down' when he wanted to make notes, a kind of bird-like foraging movement. The sense of words coming as spontaneously as cries is reinforced in many of his poems, where he darts from side to side, from plant to hill to weather to memory, containing all his experience in one simultaneous, present-moment, panoramic sweep – what Seamus Heaney called the 'one-thing-after-anotherness of the world':

> When jumping time away on old cross berry way
> & eating awes like sugar plumbs ere they had lost the may
> & skipping like a leveret before the peep of day
> On the rolly polly up & downs of pleasant swordy well
> When in round oaks narrow lane as the south got black again
> We sought the hollow ash that was shelter from the rain
> With our pockets full of peas we had stolen from the grain
> How delicious was the dinner time on such a showry day
> O words are poor recipts for what time hath stole away
> The ancient pulpit trees and the play

But Clare was as much meticulous bird-watcher as bird, not some naïf with the gift of automatic writing. These lines from 'Remembrances' are a reconstructed and carefully edited vision of Clare's experiences as a child before Enclosure changed his home patch for ever. This epic and passionate battle anthem on behalf of all marginalised lives was a piece of crafted poetry.

Which might make for a neat closure were it not for the fact that many birds also 'craft' their singing, learn from their elders and other birds, and that the nightingales which Clare so exquisitely championed are more cultured singers in June than they are in April . . .

*

Most of the late autumn and early winter I shared days in the house with Ian, one of the building team. He was officially doing the paintwork but as he worked his way quietly and methodically through the house, he'd turn his hand to any skilled job that needed doing. He combed builders' yards (and our barns) for old planks and timbers to make sink surrounds. He made the labyrinthine TV aerial system work. He had his say (and sometimes, I suspect, his veto) over the colour schemes of the rooms. One morning he showed me the layers of limewash that were uncovered as he sanded down the walls. They were in bright pastel shades of pink and green and blue, and the overlapping patches they made had the look of tortoiseshell, or the mottles of colour you sometimes find on the inside of hollow yews. Underneath was the plaster itself, laid across the oak frame and sweet-chestnut laths, and made from lime, sand, water and animal hair. Ours was horse, but anything would do. Gilbert White used his dog's when he was plastering his ceiling in the 1780s: 'His coat weighed four ounces. The N.E. wind makes Rover shrink.'

I asked Ian if he had ever done classic wattle and daub. He was young and had a trace of East Anglian reticence, but yes, of course he had. But he didn't necessarily use the usual base of split hazel. He would cut whatever twigs and poles happened to be available on the site. The one wood that mustn't ever be used – though it must have been so temptingly available in the damp environment – was willow, which could take root in the moist daub and sprout clean into the house.

Ian had an instinct for the indigenous and vernacular, and a strong personal aesthetic about restoration. He believed its chief purpose was to make a house liveable, according to the needs of the time. Of course you should make the place damp-proof, stop rot, replace irreparably broken and decayed wood, always with materials and styles in keeping with the spirit of the place. But hiding simple wear and tear, or faking antique features, went beyond organic restoration into redevelopment. Here many of the beams had been stripped of their centuries-old encrustations of wax and paint. But they were still full of nail-holes, spiders' nests and streaks of misdirected plaster. The floorboards were so ingrained with living that they would barely take a shine, and ancient beer stains did passable imitations of expensive burring. This is the 'character' of the house, not in the sense used by interior designers or heritage preservationists, but because these fabrics embody the memories of the house and become, in a way, its sensory system.

Historians are fond of using the word 'palimpsest', borrowed from calligraphy, to describe artefacts and landscapes (and maybe lives) in which layers have accreted, one upon the other. I don't think it is the word they are really after – though in the modern world it may be accidentally and depressingly exact – because a palimpsest is properly a manuscript in which one layer of writing has been erased or obliterated to make way for the next. The densest, richest and usually (but not always) most ancient landscapes are the opposite of this. The earlier writing, so to speak, shows through, or in some way influences what comes later. Curves in the land, seams of dampness, an acid bite in the soil, are tenacious things. They hold their ground. The wild inhabitants of a place show the same kind of obstinacy, clinging on for as long as they are able to the last patches of their aboriginal habitats and traditional route-ways, even as they are sliced up by new roads and 'reclamation'. (How can we use that presumptuous word, as if we were snatching back from nature something that was once ours?) Their cause is, alas, usually a lost one. But humans can also add benign touches to the landscape, and these too can resonate long after their original functions have disappeared and new forms and uses been grafted onto them. The whole of central East Anglia is, in a sense, a ghost of this kind of layered landscape, a reminder of something once more substantial. The elms are dead, the hedges gone, the fields drained. Yet the anciently massive boundary ditches, and the wide grass verges that once bordered droveways, are still there, a skeleton whose natural structural logic won't go away.

In prehistoric cave paintings in France, dating from thirty thousand years ago, new sketches were sometimes done between, over and around older pictures, horses over bears say, despite the painters having perfectly adequate techniques for covering up old work. The word 'patina', with its sense of encrustations and natural wear, comes closer to describing the end-product of all these kinds of process. But maybe a better metaphor is just 'weathering', the ongoing signature of natural forces and human work – of, simply, experience.

*

Then, abruptly, the real weather, which I'd tidied harmlessly away in the grain of the beams and a surfeit of metaphors, decided to come inside in person. Just a few days before Hallowe'en and the old festival of Samhain, when shivering East Anglians once lit fires to defeat the powers of evil and stave off the lethargies of the dead months to come, winter exploded in the valley. Brewed up in the south-west, a major gale hit Norfolk. It raked through the countryside, snapping oak trees off half-way up their trunks, and downing telephone and electricity wires. Many houses in the valley were cut off for a week. Ours escaped, but the wind was palpable inside. An hour or so after it began, a strange miasma began to drift into the rooms through the beam-joints and unsealed doors, an aerial flotsam of plaster dust, rotten wood crumbs, feathers, horsehair and fragments of bird droppings. I wondered if this might be four-hundred-year-old swift guano, sucked up in the loft and whirled down through the floors. The smell that came with this fossil dust was primordial: a sweet and dangerous odour of must and soot and dead things. The pall hung in the air for hours and then slowly settled on every flat surface, a brief reminder of old lives, but a sharp intrusion into my new existence.

The Great Blow was just a beginning. A few days later the rains began, and barely stopped for the next three months. The wet was first depressing and then absurd. It was finding its way into every conceivable hollow, filling those it had created millennia ago, and opening up new possibilities as insidiously and capriciously as a Chinese whisper. Water conspired in previously invisible dips in the lanes, in moats that had looked no more than ditches, in fields that farmers thought they had drained into submission. A film of the valley virtually underwater was shown on the news, and anxious friends phoned me up to see if we were marooned. (We were fine: the house had been sensibly built 80 feet above river level.)

Once, when I was making a film about the limestone country of the Yorkshire Dales, the sound engineer succeeded in assembling fifty-four different recordings of moving water. He'd caught the roar of fountains bursting through hillsides after heavy rain, the insidious drip of limewater from stalactites, waterfalls plunging over crags and tumbling more gently down the inside of pot-holes, and, most hauntingly, the distant, ever-present murmur of underground streams, wearing away the stone. With no rocks and little elevation, we have no such water-music in Norfolk.

But we do have water-colours. There were transformations and optical illusions conjured up everywhere by the wet surfaces. Beyond the fen at the foot of our meadow there was a flooded pasture so pale with reflected sky that, looking at it through the trees, I took it for a new crop, or a vast sheet of horticultural polythene. In the mere which the adjacent field had once been, a single swan lifted up, ghostly white, and had to weave through the tops of the willows like an owl. The farm ponds were stained yellow by the sand washed out of the fields. Almost everything else above ground seemed to have been darkened by the wet. The sheep were a dirty, disgruntled grey. The air was like damp flannel. The trees were in silhouette – except those split open by the October gale, whose sodden gashes had turned red-raw.

No wonder that East Anglia has proved such a magnet for landscape painters. The shine from the water is another source of light. From the Norwich School's fords and cattle wallows, through Constable's canals and the pond in front of Willy Lott's cottage, to the myriad waterscapes of weekend painters on the coast and in the Broads, East Anglian art is permeated by two sources of energy: one in the sky and the other in the shimmering wet.

So that insidious penetration of water in November was nothing exceptional. Norfolk is a wet county, certainly every bit as wet as its other fabled quality, flat. There may not be much variety in the gentle swells that pass for slopes and hillocks here, but this is an inverted landscape, riddled and convoluted with concavities. In Breckland in the west, there are shoals of small ponds, each surrounded by a miniature embankment. They're known as 'pingos', and were created when ice-boulders ('lenses') were carried south by the glaciers, lumbered to a halt and slowly melted into the sand. It's not really surprising that human diggings are scarcely distinguishable. Down in the valley fens there are also shoals of small pools with raised lips formed when one-time peat-pits were flooded. Water, for all its fabulous variety here, is a great leveller.

*

Today the valley has become a waterscape in three dimensions – four if you count the horizontal rain. The wetness is so all-pervading that it's verging on the comic. The drive is flooded. The lane is little more than a series of lagoons. Tidemarks, I swear, are creeping up the legs of chairs. Outside it's an astonishing sight. It's like a second spring, and the water seems to be quickening the place, pulling strings, jerking earth and wood (and me) back into action. The creeks and dykes alongside the lane are full of raging bournes. I find I'm aquaplaning down the road, and the car's screen is framing blizzards of waterspray and windblown redwings. I surf towards the big fen that lies about a mile west of the house. The peat is so saturated that it oozes like a sponge with every step. I put up a fox from a sedge-bed. It's a big bedraggled animal, and it skitters and twists from one less-wet patch to another, throwing up little jets of water around its feet. It stops behind a tussock about 50 yards away and glares back at me. Our brief relationship is not exactly harmonious. I'm elated and it is peevish. But I think I know something of what it's feeling. Too much of this and I'd become a morose, failed amphibian myself.

But a few days later I was on my first trip to the Broads, just about the most thoroughly inundated part of the whole region. It was the kind of day the Broads were made for – or made *from*, if you believed your eyes. The sun was invisible, no more than a hint of lightening in the south-east, and the mist had the kind of opalescence a trace of chalk gives to water. The scene in front of me – cracked willow pollards, meadows glassy with standing water, motionless and amorphous cattle – seemed to be settling out, condensing from the mist. It was another trick of the light, but a true enough portrait of this countryside's elemental weaving of water and sky. I was far over in the east of Broadland, at Horsey Mere, and began following the track that winds round the edge of the water. I nudged my way through head-high reed and birch scrub, sending flocks of bumbarrels – the delicious local name for long-tailed tits, inspired by their tiny rotund nests – flitting from tree to tree. I caught occasional glimpses of seemingly identical patches of water, and my sense of direction began to go flabby. A thatched boathouse swam into view on one side of me, then reappeared on the other. Overhead, skeins of pink-feet geese flew in and out of the mist. There were woodpeckers everywhere, attracted, I guess, by the dying alders and willows, but heightening the uncertainty about whether this was a wood or a wetland. In this shifting, quaking place, the one very quickly turns into the other.

In the afternoon the sun shone through and the mist evaporated. I wandered back along a lane between the Mere and the sea. A marsh harrier lifted up from a reed-stand and drifted to a dead tree, an effortless, oriental slide, a shifting of air not body. Landed and out of its element it looked bulky and awkward, too big for the branches. Then, out of the corner of my eye, I glimpsed a small group of large, long-winged birds putting up near a dyke. I got my glasses up quickly, and for a few seconds made out the unmistakable silver-grey plumage and swan necks of three cranes. I'd only seen cranes once before, in huge numbers down in southern Spain, where most of the western European population spend the winter.

In the Broads they're a new treasure, glamorous, cryptic, talked of in confidential terms. They're prodigals, too. Cranes bred in the Broads back in the sixteenth century, and probably throughout the wetter regions of East Anglia. For the four succeeding centuries they only ever appeared as passage birds, blown off course on their migrations between Scandinavia and the south. Then, in September 1979, three turned up in the Broads and, unprecedentedly, stayed for the winter – and for the succeeding summer. They raised their first chick in 1982, were joined by other vagrant adults, and by 2003 had become a slowly expanding colony of between fourteen and eighteen birds.

Cranes are the epitome of wild places, and their return to the Broads is a blessing, a sign that their sense of wilderness hasn' been entirely destroyed. They had come of their own acc settled where they wanted, and survived without any elaborate protection systems or habitat manipulation. No wonder that across the world, wherever cranes breed, winter, or just pass through, they are regarded as symbols of good luck, renewal and fertility. Their migration flights are epics – the birds etched against the sky, their faint, excited trumpetings rippling through the flock. When they reach their breeding-grounds they begin their extraordinary festive dance-displays. At Lake Hornborgasjon in Sweden, where fifty thousand people gather to cheer them in each spring, many join the birds in their dance. When I'd watched cranes in their winter quarters in Extremadura, they were sauntering about in small family groups, feeding off acorns from the cork and holm oaks. Yet even this utilitarian business was a kind of dance, an elegant slow-footed gavotte.

My new neighbour and fellow writer Mark Cocker, a watcher with almost superhuman senses, had seen them dance in Norfolk, and I wanted to, badly. As the weather worsened I yearned to tune into some ancestral rite, some spell to deter the insidious onset of winter. But the fog came down, at just the moment East Anglian farmers began taking in the sugar-beet, and turned the harvest of this noxious crop into high drama. The immense machines, as tall as houses, worked on into the darkness, with racks of headlights glowing palely through the mist. From my study window it looked like a scene from *Close Encounters*. The beet-stack next to the farmhouse swelled until it was more like a wartime fortification than a pile of vegetables. The next day the harvesters were joined by tractor-drawn ploughs and harrows and drills, until there were often three different machines following each other round the fields. Hares scattered towards the hedges and fugitive pheasants lined up on our wall. Within two weeks the winter wheat was already a mist of green over the bare soil.

Then the temperature dropped and the wind veered round

to the east. Norfolk's notoriously cold blast comes straight from the steppes and the gusts that six weeks before had wafted in a delicate dust from the seventeenth century were now blowing in what seemed like vaporised permafrost. The wind burst in through the joints and window cracks. It held the cat-flap open in a horizontal position. I could feel it on my face at night. The house was hyperventilating, gasping in cold air and exhaling warm through every orifice. Outside there was a chill factor of −10 degrees, and in my room the thermometer barely reached freezing point. I couldn't work. I stuffed old pillows up the chimney and rolled-up bin-liners into the biggest floor holes. I went round every window seal with draught excluders. In desperation I took my typewriter down to the kitchen. The cats had had enough too. They curled up like snails on the warmest surfaces they could find and went, sensibly, into temporary hibernation.

Outside it was survival time. One of the beet-field hares lay in the road outside the house, shot through the neck and with its guts picked out by crows. Flocks of golden plover appeared in the ploughlands, escaping here from the frozen Russian tundra. They fed in mixed gatherings with lapwing, but were restless, urgent birds, darting about in sharp, tight bands. Pheasants tried to walk with their tails blown at right angles to their bodies, and then took to bickering. I'd watch cocks square up to one another, make little runs forward, jump balletically into the air, exchange a few pecks, and run away. Indoors, Kate and I bickered, too. Back from London to find a nearly empty oil-tank, she berated me for my lack of suitable winter clothing and for wearing bare feet round the house. I retorted, sulkily, that I did not want my sustaining home-made bread to share quarters with her sliced white. The central heating control swung on and off as, pheasant-like, we sparred for position.

One day when I was out, Shaun the head builder brought a

dead monkjack to the house, strung it up in the apple tree by the guinea-pigs' hutch, gutted and butchered it, and stuffed the joints in Kate's deep-freeze. Later, I got an expanded version of the story from him. There had been two deer crossing the road and he'd hit one badly. He thought of our looming Christmas, cut its throat and lugged it into the back of his van. He hung it for nine hours, so that the blood either drained out, or (he knew his physiology) 'into the pleural cavity, so it would come out with the innards'. He then cut off the forelegs, carved out the hind legs, shoulder, saddle and fillets, and put them all separately into plastic bags. I asked nervously (he was a big man and omniscient about all local matters from weekenders' gossip to the ancestral migrations of water) how he learned to butcher so well. He'd started when he was eleven, he said. Dad had a smallholding, and he'd learned on pigeons, rabbits and the odd pheasant. Also, he added cryptically, 'a few pigs came our way'. Fortunately, the hapless monkjack wasn't the one that strolled about our meadow in the early morning, but its clotted blood lingered in the grass below the apple tree until the snow came.

The French use the same word, *temps,* for time and weather, suggesting all kinds of subtle linkages between the two. Ecosystems evolve in time precisely because of weather. Weathering is an experience which permeates the lives of all creatures, and important moments or events – a storm, a migration, a nervous breakdown – are partly expressions, *freezings*, of environmental conditions in the broadest sense.

I'd just learned another useful foreign term. Up here they call a parcel of rough weather a 'blonk' – which seems perfectly to capture the feeling of being hit broadside by a bad spell. I was definitely blonked, and beginning to sink under-winter. I knew that weather is the one part of nature we can't master, and that I ought to relish it in all its moods, like Coleridge. In a poem for his son Hartley, 'Frost at Midnight', he wrote:

> Therefore all seasons shall be sweet to thee,
> Whether the summer clothes the general earth
> With greenness . . .
> Or if the secret ministry of frost
> Shall hang them up in silent icicles . . .

There were no icicles yet but the cold and louring grey skies were getting to me. Their stolid refusal to shift seemed to be a denial of all the quickening and hopefulness I was beginning to absorb from this landscape. I could enjoy the wet but not the dark. Years ago, at this time, I used to look forward avidly to 12 December, the day when there is the first chink in the murky procession of days. Because the earth's orbit round the sun isn't symmetrical, the relentless pincer movement of darkening mornings and afternoons doesn't conclude with a neat snap on Solstice Day, 21 December. The mornings don't begin to brighten until New Year's Day, but the evenings 'draw out' from the 12th. I'd clung on to this notional glimmer of new light and always tried to celebrate it with friends.

But this year 12 December dawned claggy, cold and overcast. I was feeling cut off and morbid. There was nothing unusual about this in bleak winter spells – except this is how I was feeling at the beginning of my illness, three years before, and it was unnerving me.

Illness is the dark side of our transactions with nature. It's a reminder of the routineness of death, of the disposability of individuals, of the fact that living systems can be ruthless and unpredictable in their constant manoevring. But, at first sight, depression doesn't fit into even this austere picture. There's no random physical 'accident' behind it and nothing which benefits, no opportunist virus or evolutionary climber. It seems to have no connection with the biological business of living at all. And what it did to me was unearthly, in that it negated, cut dead, all the things in which I most believed: the importance of sensual engagement with the world, the link between feeling and intelligence, the inseparability of nature and culture. And it began to grow at the most unexpected time, when by all conventional psychological theories, I should have been awash with the sense of well-being that comes from high status and achievement.

I'd just come to the end of the most demanding book I'd ever worked on, *Flora Britannica*, and it had gone down well. Even the hard graft of writing – a quarter of a million words on the folklore of the British flora – had been a joy, if an exhausting one. 'Folklore' doesn't do justice to the texture of the project. It wasn't quaint and antiquarian, but a documentary record from the grass-roots. Close on ten thousand people had contributed stories of the meanings which wild plants held for them, of plants pursued, nicknamed, eaten, woven, carved, hugged, dreamed about; plants used in seasonal ceremonies, home-brewed nostrums, children's games, lucky charms; plants which marked or symbolised territorial boundaries, births, childhoods and deaths. I'd added my ha'p'orth in a commentary which linked this great communal narrative, filling in some of the social and cultural history, fragments of ecology, glosses on the contributors' stories, and some personal aesthetic impressions. I drained into it – in a way I hoped didn't appropriate it from the real authors – everything I'd ever known and felt about plants, and probably nature too. I wondered what else I would ever have to say, and if the role of archivist was my destiny, now fulfilled. I hadn't then thought much about, so to speak, the nature of nature-writing, and about where on earth such a marginal business fitted into the swirling currents of real life that were its raw material. That winter, after the dust from the book had settled, a little bit of me wished I had been the conker-player, not the census-taker.

Something wasn't quite right, and with hindsight one particular experience, of no significance at the time, seemed to symbolise what it was. I'd been out in the Thames Valley, not far from my old home in the Chilterns, searching for remnants of the huge drifts of snake's-head fritillaries that had grown there up till the 1950s. This extraordinary, sultry flower had been so plentiful in meadows around the village of Ford that they used to hold a 'Fritillary Sunday' there in May, when people could come and pick bunches for a few pence in the charity box. On May Day itself local children played trick-or-treat with exotic wands made from snake's-heads and crown imperials.

The meadows were ploughed up in the early 1950s. But I had a hunch that a few plants might have survived in damp field corners, and that April afternoon I found three blooms on a small patch of common meadowland that had, short-sightedly, been planted with trees. I was elated, and strode off down the lane swinging my binoculars and notebook. I must have looked for all the world like the man from the planning department on one of his usual wrecking missions. Two elderly locals were leaning against a gate and, in the most courteous way possible, asked what I was about. I felt an odd mixture of embarrassment and curiosity, and asked what I suspect were patronising questions about the plants. They quickly put me straight. They remembered the meadows, and the great May fritillary festivals, in vivid detail. Several villagers had taken corms before the meadows were destroyed, and one of their neighbour's colonies had increased from 25 to 250. I asked if this was in his garden, but was parried with the diplomatic answer 'No, in a piece of ground.'

I went off feeling not so much put in my place (which I had

been) but put somewhere which wasn't a place at all, into a kind of anchorless bathysphere from which I peered for glints of light. I adored writing, and felt that it was important, but had a growing feeling that I was an outsider, an ecological voyeur. And it was that nagging sense of ill-fittedness that took root in me that winter after publication.

For the first time since I'd been a full-time writer, I'd come to the end of a book without a new idea straining in the traps. That hackneyed, end-of-the-line question began to form in my mind: Is That It? For years, I'd used work to patch over and manage the gaps where my life didn't quite fit, the failed relationships, the solitariness (fiercely protected and despaired of at the same time), the stubborn, or fearful, rootedness in the family home. I'd found writing a compulsive calling, if not exactly a role. And I'd had another kind of calling – helping to look after my mother when she was invalided with Parkinson's Disease. It was a responsibility that it would never have occurred to me to shirk, but also, I fully recognised, a cast-iron excuse for not getting out of that house and out of my emotional rut. After my mother died, I nursed that big book instead. Now they were both in the past. My cover was blown.

On cue, my body decided to protest. I began to fall prey to a florid array of psychosomatic symptoms. My limbs, guts and bladder ached. I began having heart arrhythmias, the 'skipped beats' known as atrial ectopics. One spring day I marched with clenched teeth round the fields with a portable ECG machine strapped to my chest, and notched up a record three and a half thousand missed beats in twenty-four hours. My consultant said they were a spectacular recital but entirely benign, and that I ought to regard the fact that they hadn't thrown my heart into further disarray as a sign of my robustness. But since I'd experienced every jolting thump, this seemed like cold comfort.

None of this was new. All my life I'd been prone to spells of malaise at times of loss or disappointment. Sometimes bad weather was enough to spark them off, or then again a bad mood that soured good weather. In the journals that I've kept sporadically, my chirpy entries on the arrival of spring migrants and balmy autumns are repeatedly interrupted by complaints about my internal climate. Early spring (April can indeed be 'the cruellest month') and early winter were often the worst. Sometimes the symptom clusters used to occur on the same days each year, like those recurrent spells of weather known as Buchan's Periods. When I was a child they were attributed to my being 'highly strung', like some corporeal wind-harp, prone to plaintive wailing when the emotional weather was bad. I suppose that these days I would have been diagnosed as suffering from separation anxiety or rejection sensitivity, from some quirk of genetics or upbringing. But what I learned was the power of psychosomatic illness – to grab attention, rebuke loved ones, to get you out of unpleasant situations, get you *home*. In the way they convey reproof and pitiableness at the same time ('look what you've done to me'), psychosomatic symptoms are simply elaborate versions of childhood sulking or tantrums. They are part of the theatre of the body.

But this time, it looked more serious. The symptoms didn't vanish with the end of winter. Over the next year or so they kept returning in new guises and combinations. And, of course, the more I worried about them the worse they got. I began to think morbidly about my future as some kind of cripple, and sank imperceptibly into genuine depression. I stopped having ideas. I lost my taste and hunger for work. I gave up going out into the country, effectively cutting myself off from the main source of stimulus in my life. I got treatment but it scarcely made any difference. The drugs didn't work, and the talk therapy – though diverting and convivial at times – soon became not much more than a ritual rehearsal of my now familiar family dysfunctions, and the possible influence or otherwise of my alcoholic father and subdued mother. Talk therapy can be comforting and a crucial point of contact with the world. But the idea that discussing or simply understanding an illness will in some way make the hard-wiring that caused it disappear is, as most people who have been through it acknowledge, wishful thinking.

So I sank further. I couldn't bring myself to do anything, so overpowering was the anxiety brought about both by choosing and not choosing. My life fell into a pattern in which I could barely distinguish one day from another. I stayed in bed most of the day, quaking with anxiety about the anxiety I was trapped in. At lunchtime I crept a hundred yards to the nearest pub, drank too much, read their copy of *The Times* from cover to cover without taking in a word, and then crept back to bed again. I stopped bothering to answer the phone or open the post.

To say that I was suffering from loss of self-esteem (a common theory about the causes and nature of depression) would have been hugely to inflate the sophistication of my feelings. My condition was the only focus of my thought. It was like being trapped in that mathematical device called a Möbius Strip, a circle of material with a single twist in it so that it has only one surface. Anxiety was the surface of my closed strip. I had one ephemeral moment of peace each day, a few seconds of equilibrium between waking in the small hours and the panic kicking in. It was an interval as brief but palpable as going through neutral during a gear change, but it kept me aware there was a state called 'well', and that it was that I wanted to be in, not out of it altogether. The lure of the peg on the back of the door was never very strong (who would have found me?), but my escapist fantasy, when I'd drunk enough to soothe myself a little, was to go feral. When my money ran out I'd use the one survival skill I did have, take to the woods and live by foraging. But I knew perfectly well it was a ludicrous dream, and mostly I just lay there, curled up in the 'crash position' they demonstrate to you on aircraft, scared stiff and praying hopelessly that the turbulence would pass.

The triggers of what was happening to me are clear enough now. I was literally exhausted, played out, up at the end of a cul-de-sac. I'd failed to go through a crucial stage of development, had never 'fledged'. But I wish I'd been in a state to think calmly about why my nervous system should have chosen this bizarre and counter-productive response, and about what might have been its original survival value. A dysfunction as common and pervasive as depression must have some kind of precursor in 'normal' behaviour, some aboriginal usefulness, some ecological role. One evolutionary psychologist has suggested that its prototype is 'the feeling of the hunter who returns home without his prey', which encourages him to be more successful in future. This sounds like the template for male pique rather than prostrating illness. Orthodox psychology has long been dominated by hunting-male scenarios, and the 'fight or flight' model of reaction to stress.

But throughout the living world there has always been a third response to trouble, the stratagem Oliver Sacks has referred to as 'vegetative retreat'. When fight or flight is impossible or inappropriate, or maybe not properly learned, organisms lie doggo. Possums 'play possum'. Hedgehogs curl up. Barn owls faint. They all pass, for self-defence, into states resembling death or sleep, where decision-making and even movement are put on hold. Even nestling swifts do it (admittedly for different reasons), going into a state of torpor when their parents are away for long periods on feeding expeditions. Vegetative retreat means safe harbourage, a period when inward, protective processes take precedence over all that adrenalin-pumping. It's an entirely sensible response by any creature to a threat which stumps it, a kind of ur-depression. In humans, prolonged unhappiness or disappointment also seems to provoke it. In nature it's meant to be a short-term tactic, and to dissolve when danger passes. But mine didn't, and I clearly thought some unspecified danger was still there.

So I put myself in hospital, in the hope of breaking the vicious circle. I was dried out and looked after. From inside the patients' recreation room I watched a hobby hunting over the hospital grounds, and felt pleased that I could still recognise it. I enrolled for occupational therapy. But all tools and interesting materials were banned. So I did a few days of calligraphy, enjoying briefly the joke of teaching myself to write exquisitely.

And it dawned on me, just before I was discharged, that I was in the same hospital in which John Clare had spent the last twenty-three years of his life. At that point I didn't know much about Clare's illness, or I would have begun fantasising similarities between our fates. He'd been a 'highly strung' and hypochondriacal child like me, racked by inexplicable pains and a deranged digestion. His distress had increased as he aged, aggravated by the sense of loss and disorientation that the Enclosure of his home landscape brought, by poverty, and by the racking demands of his publishers and patrons. He was, as the best-known 'peasant-poet', a displaced person. But he was almost certainly a manic-depressive, too. In his forties he had become sufficiently disturbed and delusional to be admitted first to an asylum in Epping Forest (from which he escaped and walked home), and then to this hospital in Northampton. Dr Skrimshire, who filled in his admission papers, put down the chief aggravation of his condition as 'years addicted to poetical prosing'. But he was well looked after, as I was, by sympathetic doctors who encouraged him to continue writing. He was allowed into the town in the daytime, and became the town versifier, sitting in a church porch and composing birthday rhymes and billets-doux for anyone who would give him a screw of tobacco. He continued to write more serious verse – some of it heart-rending, some pure doggerel – until a few years before his death. He was, in the words of Jonathan Bate, who has lovingly and exhaustively documented his life of ecstasy and struggle, 'heroic'. Yet there was a perceptible and probably predictable change in his poetry. During his asylum years Clare was never in the remotest sense 'vegetative' in his depression. But he did go progressively into retreat, and the vivid, sensual, connected verse of his middle years was replaced by more introspective, abstract, almost metaphysical musings. Poetry itself, 'the muse', the metaphor of the lost and disappointed singer, became his themes at the end, not the real and electifying songsters with whom he'd once shared 'the best part of summer's fame'.

I went home after a couple of weeks, partially mended. But nothing in my circumstances had changed, and I started to go rapidly downhill again. I shut the curtains of the room I lay in, not just against the embarrassment of being glimpsed, but because I could not bear to see reminders of what I had lost. So I gazed instead at the bookcase, at the collection of books I'd built up over a quarter of a century, and which seemed unlikely ever to be opened again. And at my own volumes too, whose creation now seemed an incomprehensible mystery. The loss of that bit of me was the hardest to bear. Writing wasn't, melodramatically, my 'reason for living'. It was simply the way I saw things. To have an experience was to make it into a phrase, or a scene, or a story. When my father died, the first serious event of my life (I was twenty years old), all I felt at first was relief that a remote and tryannical drunk had been lifted from our family's life. But at his funeral I was seized by a desire to make amends, find some fondness for him, at least lose myself in mourning. It didn't work. The scribbler on the shoulder took control, and when I got home from the graveside all I could do was shut myself in my room and write for hours, about the shoppers who doffed their hats to the hearse, and the strange shininess of the grave-earth, and the moment my long-suffering mother had almost toppled forward. To lose that reflex was like losing the instinct to put one foot in front of the other.

And now, deprived of outside stimulation, my senses turned in on themselves. My ears, already a little noisy from high-tone hearing loss, began to improvise. I started to hear things – not hallucinations, because I knew perfectly well they were only inside my head – but with great clarity none the less. In my left ear was a four-piece brass band, playing what is usually called 'easy listening'. In my right, a Russian Orthodox bass, with a voice of extraordinary range and power. I could vary the brass band's repertoire at will (though it seemed, mischievously, to have a special liking for 'The Floral Dance') but the Russian bass remained uncontrollably wild and florid. Then, one morning, I woke up terrified by all the red book-jackets in the room, which seemed to have taken on a sinister glow. My doctor, trying to humour me I suspect, said that there was more blood in the retina first thing in the morning, and if I was going to find something to be anxious about that would be first in the frame. But I was alarmed that my own blood-flow could scare me, and wasn't about to forget that red is nature's all-purpose alert signal.

The last few acts can be quickly told. They're the familiar final stages of a breakdown. The unopened bills turned into debt-collectors on the doorstep. I abandoned taking care of myself. My sister, with whom I'd shared the house for most of our lives, found it impossible to deal or talk with me any longer, and was insisting that we sell the house. This *was* it. So I heeded that red warning, put my affairs in the hands of some old friends, and took out the last of my depleted savings to pay for a few more days in hospital. I was dried out again and stuffed full of the maximum possible doses of a battery of drugs. But there was a difference. This time neither my doctors nor my new attorneys would countenance my discharge if I was to go straight back home again. I must either go to a half-way house – something which would have surely finished me off – or into the care of some reliable friends, who by now had mercifully realised the state I was in. (That was thanks to Di Brierley, who had visited me one especially dark day, and had the nous to go through my address book.)

So I was packed off, straight from the hospital, to the north Norfolk coast, where I'd first gingerly poked my toes beyond the cosseting of home back in the 1960s. Mike and Pooh Curtis, who'd been friends since I first discovered this marshland edge, took me in hand like a foundling. I was weighed, kitted out with new trousers to accommodate the weight I'd put on (but left in no doubt that I had better get it off) and set to work. I hung out washing, picked vegetables and went shopping in the village, an exercise so unaccustomed that I had to find hiding-places to sit down and rest. Pooh gave me a notepad, just in case a thought might come into my head. It didn't. If my illness had been a vegetative retreat, this was a kind of vegetative advance, a slow, grinding, mindless pull back to some semblance of self-sustaining behaviour.

Within a week, they were testing me on the salt-marshes about which I'd once written so longingly. 'The closeness of these unstable edges of the land', I'd mused, 'was part of the secret of Norfolk's appeal . . . a reflection of a half-conscious desire to be as contingent as spindrift ourselves, to stay loose, cast off, be washed up somewhere unexpected. Down among these shifting sands the world seemed full of possibility.' Now their instability looked the most threatening thing in the world. I was taken cockling, then samphire-picking on the flats. I used to find this a deeply satisfying and exhilarating business, wading ankle-deep in the mud, bunching up the succulent shoots, marching home like a hunter-gatherer. But now I couldn't squat down for more than twenty seconds without excruciating pains in my back and knees. Not many years before I'd been able to run barefoot over the mud, quite sure of where to put my feet. Now my ankles, unused to carrying weight for any length of time, went over every quarter of a mile, and I was left floundering and limping. When I went on my first serious walk — three-quarters of a mile — I ended up barely able to see or breathe. Buzzed across Blakeney harbour in Mike's small boat, I fervently wished I was back in bed with my face to the wall. The waves lapped inches from my seat. I felt too stiff and scared to climb in or out. I denied, petulantly and falsely, that I had ever felt any affinity with the sea. Pooh simply shook her head and said, 'If you're coming to Norfolk, Rick, then you'll bloody well have to learn to deal with water.' She did not know quite how prophetic she was being.

But slowly and laboriously, I began to improve. My weight dropped off and I started to breathe properly again. I wrote a postcard. I noticed the smell of broad beans. Playing the part of a well person, pretending that there might be a future, made it happen, little by little. The performance became a shell — an embodiment — in which feeling could begin to happen again. Or, then again, perhaps it was like the way in which the workings of a watch can be re-kindled just by winding the hands.

And so the weeks progressed. I was passed from friend to friend, along the Norfolk coast to Justin, who'd first introduced me to Norfolk as a teenager, and was now my finanical keeper; inland to Roger; down south to Di, who had first rallied my friends; then, daringly for a while, in my old home country (though not yet 'home' again) with Francesca and John, who had been my colleagues in Hardings Wood; then back again to north Norfolk. From feeling like a helpless piece of flotsam, I began to rather relish this entirely care- and responsibility-free life. I barely gave a thought to the fact that in a few weeks, I would no longer have a home.

I'd met Polly about eight years before, at a Norfolk dinner party. We'd been placed next to each other, from our host's kindly belief that a shared love of plants would give us something to talk about. It did. She was helping to make a facsimile of a Benedictine herb garden in the heart of Norwich, and wanted to know everything I could tell her about such things. She had that kind of arresting curiosity and intense attentiveness that is both flattering and deeply attractive. She was also a Norfolk girl, brought up as a doctor's daughter in the Broads, with the same kind of wild childhood I'd had in the Chilterns. Now she was a lecturer in childhood studies, and had got herself an art history degree in her spare time. We had indeed a lot to talk about, and the three hours passed with us barely speaking to another soul. I don't remember any of the details, but I do remember the way we spoke to each other, and the facial expressions we shared of hugely exaggerated surprise and excitement and scepticism – mouths gaping and gurning all over the place, in the way that children do almost as soon as they learn to talk to each other. At some quite deep level we had already recognised the unreconstructed hedge-kid in each other. But she was settled with a husband and children, and at the end of the evening I put it aside, ruefully, as just another one of those 'what if' encounters.

I saw her again now and then when I was in Norfolk, and the meetings were always as pleasant and poignant as that first time. Now, back on the north coast in rather different circumstances, I wondered how she was, and if we might say hello again. When I was told that things were far from all right in her life, I persuaded my hosts to invite her for tea. She came to visit us a couple of days later. We walked (chaperoned) over the fields, looked intently at a lot of flowers and more nervously at each other. Later that week she showed me round her herb garden. We flirted gingerly, riskily. I picked up the unseasonal scents of lily-of-the-valley among the fading autumn perennials. 'It's me,' she blurted. We stole a kiss, eyes wide open. A couple of days later, while I was staying at Roger's house in Mellis, I swallowed hard and asked her if she would like to come and see me. I waited for her outside the pub on the common. She arrived half-hanging out the car, waving like the queen. We stood in each other's arms for so long, amazed at what was happening, that passers-by began to shout ribald jokes at us. The afternoon was clear and hot, and we walked through the harvest fields. It was confessional time, the laying of losses on the table. She told me about the sad decline of her marriage, and about the death of her brother, a wound still not healed. I told her of my own illness, my mother's longer illness, the lost years. We cried a lot, the gasping tears of release. The sky was an astonishing blue. When we lay down we could hear mice scrabbling through the wheat. We discovered how corn circles were made, and laughed, at last, when it all turned into a tangled comedy. We left the field, stubble-scarred and knowing that both our lives had changed.

I was well enough now to risk going back to the Chilterns, for the passing of our family house. My money had almost run out, and I wrote a self-pitying note to Polly, complaining that I did not know where my next meal was coming from. She replied by return of post, with a Jiffy bag full of Swiss chard. She made it very plain she was not about to mother me, and that she already knew instinctively what would put me right, and make me laugh. I doubt that any one else in the world could have guessed the meaning that a bag of leaves would have for me at that moment. It was only by lucky chance that I found the small envelope, tucked among the greenery. In it was a bank-note, not for food, but for the train-fare to London, to an exhibition at the Royal Academy. We both played truant and met there. I spent the next hours in thrall, to Polly's view of art not just as an end-product, but as an act of *making*, a drama that involved the artist in a moment of living, with moods, friends, weather and bills, as well as vision.

She had, at the tail-end of the summer, to go on a cycling holiday over the Pyrenees. It seemed impossibly far away, and, to both us, a very painful absence so early in our relationship. But, acting very grown-up for once, Polly asked me to keep a little diary while she was away, so that she could know what I had done and thought. I do not know, and hope never to be told, just how much of a wise therapeutic hunch there was behind that little lover's whim. But I did as I was asked, very gladly, and for the first time in two years began to scribble words that were more than my name and address. I wrote about what was happening in the garden, about birds passing through, about my wood, about a trip out with friends to a Bank Holiday rock festival in a country pub, where I'd watched a six-year-old dancing cajun like a Louisiana native. I tried to reassemble bits of my past that I hadn't yet talked about, my years in Oxford, the jobs I'd had before I became a writer. And the more I wrote the less my life seemed to resemble that of the marginalised voyeur I'd cast myself as. I'd been places, made friends, and made, I began to remember, a mark. Writing again was not just a relief in itself, the regaining of an identity, but a key that began to unlock pieces of me that had been dormant for years. I put down on paper, I think for the first time in my life, some unashamed erotica, and, as if I was nervous of being mistaken for someone I wasn't, a complete and detailed political confession, including not only the personal ecological credo I should have written years ago, but my ownership of a map of a Regional Seat of Government during peace movement days (I'd hidden it in a matchbox in my clothes cupboard). The pages grew, and while I wrote, sitting under the shade of the beech tree that had sown itself in our garden thirty years before, prospective buyers came drifting across the lawn. Mostly they gave me a polite but nervous berth, worried perhaps that I might come with the property. They didn't know that I was, indeed, the writer in residence again.

At the end of Polly's holiday I realised I'd written what was virtually a short book, and that I had my life back again. My friends, the sea, the hard work, that symbolically comforting beech, and above all Polly's care, had broken the grip of my illness. But it was regaining that imaginative relationship with the world beyond that was my 'nature cure'. It would be with me when I packed my things, moving to East Anglia for love now as well as luck.

*

Thankfully, I was not about to relapse. The solstice came and went, a notional moment in the minute shift between darkening and lightening. And the weather, on cue for once, turned balmy. The winter wheat reached 2 inches. In the freezer, the monkjack became an embarrassing fossil. But Kate was down for Christmas, and graciously invited me to spend Christmas with her family. Having been about as convivial as Scrooge for the past two winters, and having rarely spent a Christmas with more than three people since I was twelve, I was touched. In return, I offered to help her decorate the house, only dimly aware of the scale of the tradition I was briefly joining. First, the packing cases came out, and in a short while every working surface was covered with the bric-à-brac of three generations of winter revels: a whole box of iron and china swallows, showing what seemed to me (my taste for trivia rapidly whetting) the clear characteristics of *Hirundo daurica* rather than *H. rustica*. A whole box of ducks was less taxonomically clear, but many were wrapped in parcels of three. There were wedding present plates, an assortment of tree fairies, bags of tinsel, and at least four drums of lights, all of which had to be tested, and which were to be strung not just round the traditional spruce, but up the woodpecker's pear tree too. Not wanting to be just a bemused removal assistant, I volunteered to assemble the crib, which came, heavy with age and significance, as a kind of kit: a cardboard box labelled 'Cod Pieces', a handful of straw, and some vaguely biomorphic figures made out of baked clay, which I think were the relics of a school pottery class back in the 1950s. I tried to remember my own school R.I. (about the same vintage), to distinguish shepherds from sheep, and find the primal blob that was intended to be the Christ child. For a laugh I put the wise men at the back of the fish-box, and all the animals at the front; then decided that ecological satire was not in order at this moment of the year (though it was none other than St Francis himself who began the tradition of the *presepio*, with a real ox and ass standing beside the crib).

On Christmas Eve the guests began to arrive, an entire clan network of children, children's partners, sisters, uncles, nieces, and relations. I had to draw myself a diagram to work out who was what. Everyone was kind, but I was uncomfortably aware that I was, with the best will in the world, an outsider, a little like one of those hermits who adorned the scenic backcloth at picturesque parties in the eighteenth century. Given my odd social profile – a singleton lodger with a dubious job and dodgy past – I was grateful I was given house-room at all.

Christmas Day began with the sipping of champagne, before we all sat down on pews round big tables in the newly painted dining-room. We talked about the importance of family tradition and of 'winterfests' to see off the dead season. Outside the sun blazed, and the gnat swarms danced outside the windows. Inside, we began the first of what proved to be an exhaustive schedule of rituals. Alan, Master of the Revels, organised us into a game of Trivial Pursuit, one side of the table against the other, light, as it happened, against shade. It was a bold structural opposition, and I began to wonder if we might go the whole pagan hog, with the forces of Light ranged against those of Darkness, as we charged around the leafless orchard with a flaming bush. 'Just one more round,' pronounced Alan. Lunch seemed a long way off.

*

On Christmas Day exactly seventeen years before, a field biologist from the University of Vermont was sitting on a snow-covered hillside in front of two coyote-killed sheep he'd dragged up from a nearby farm, and pondering what he'd witnessed the previous autumn. Professor Bernd Heinrich had been sharing the remains of a carcass with fifteen ravens ('There's no greater pleasure than eating roasted moose whilst resting under a spruce and contemplating ravens') when he 'saw a paradox'. The ravens were lunching equably en masse, and summoning up more birds with a 'yelling' call he hadn't heard before. He was, he confesses, awestruck. What he calls this 'left-wing co-operation', this sharing beyond the kinship barriers, seemed to go against the foundation of all his biological training, and against the holy writ of selfish-gene theory.

For the next five years he studies ravens obsessively, lovingly, despairingly, and his journal *Ravens in Winter* becomes a classic account of the experience of the cold season, of the customs of birds and the rites of biologists, and of the perennial struggle (almost constitutional in Heinrich's case) of science with passion. The riddle taunts him. Can birds willingly *share* their food? Is the deliberate 'recruitment' of other birds a fact, and if so, to what possible evolutionary purpose? He tests theories, and himself, almost to destruction. He sleeps with dead goats and cats, his raven bait, and drags 40 lb of pig entrails up a frozen mountainside. Sometimes it is so cold in his cabin that the clock slows down, and wrecks the timing of his meticulous observations. He becomes a kind of shaman to the ravens, and plays their intricate repertoire of calls back to them over an immense sound system. Soon he has nine different hypotheses on the go: incomprehensible sociobiological models like the Prisoner's Dilemma; more plausible theories, such as the possibility that the birds are seeking safety in numbers in case of predators, or conversely, that they are making a hullabaloo to *attract* predators, to open hidebound or deep-frozen carcasses for them. Meanwhile, the ravens continue with their bewitching, perplexing antics, and Heinrich falls ever more in thrall to them. 'The birds are totally at ease. After feeding, some roll on their backs in the snow like happy dogs or lie breast-down, fluttering and kicking snow. Some slide on their breasts, pushing themselves forward with their legs. They are snow bathing . . .'

Birds and watcher begin to merge. Down on the Vermont campus, the professor holds a celebration, and recruits a gang of student helpers to feast on a lamb carcass. They guzzle Moosehead beer and yell to folk guitars, and the whole shebang culminates in the Cage Raising, the building, Amish-style, of a vast raven coop where he can study the footnotes and subtleties of raven behaviour at will. ('I can't help but wonder what they would make of some of our customs', he writes wistfully.)

After five years of such devoted, affectionate, exasperating soul-searching – 'I don't like unpredictability. I want uniformity of results' – you are willing Heinrich to dump his ideology and go with his gut instincts, his spirit, his obvious empathy. In the end he makes a list of fifty-three clues and hypotheses, and just a little reluctantly, since this is what scientists must never do – accepts the obvious. The yelling, feeding mobs are juveniles, doing what all young creatures do at feast times: forming gangs for the purpose of making friends, seeking status, finding mates and having fun. You're just thankful he came to this conclusion the longest possible way round.

*

The juveniles on the Ling – the heathery common that stretched beyond the house – were also stepping up the pace. After lunch there were more ritual games and a good deal of social display. As with all family Christmases it was important that things should be done as they had always been done, and in the proper order. First there was a furious dancing game, which I couldn't make head or tail of, and then a treasure hunt I couldn't keep up with, so energetically did the younger guests hurtle round the house. I began to feel not just like an outsider but an old crock. The climax of the festivities, which loomed with an awful sense of inevitability (and the likelihood of divine retribution) was to take place round my crib, which had been tidied up for the occasion. There was to be a tableau, and I was cast as a shepherd. The regular shepherds knew the form and donned tapestry curtains for robes and tea-towels roped round their heads. The smallest cousin stood in as the baby Jesus. I did my best with bits of ethnic clothing, but the combination of Spanish shoulder-bag and violet bandana made me look like a window-dummy from a 1970s boutique. We froze, and photos were taken, pictures that would join ones taken the year before, and the year before that, and back, for all I knew, to the time the family first had a camera.

The critic Hugh Sykes Davies has coined the word 'ecolect' for the distinctive language and customs that small family or friendship groups use with each other to cement their special bonds, and reassure themselves that things are as they have always been. It is, I think, a universal phenomenon among living things. I was just feeling a bit ecolect-free myself.

*

But of course I was only as landlocked as I chose to be. Over Christmas I'd heard from David Nash, his greetings spiced with a piece of news he could never have guessed would be so timely. His wild, quixotic sculpture, *Wooden Boulder*, had reached the sea. I'd first seen this oddly touching lump of wood five years before, hunched in a stream near Blaenau-Ffestiniog, brooding on its escape. Back in 1978, David had cut down a large oak tree which overhung the house of some neighbours, and acquired the trunk as his payment. It became his first 'wood quarry', and he cut out a hulk of wood, weighing half a ton, which he faceted with a chain-saw until it was almost spherical. His original intention had been to move it to the valley floor, then cart it back to his studio. But he then had the idea of using a stream to transport it, down through a series of pools and waterfalls, so that its progress could become a kind of narrative work. *His* narrative, that is. The oak block had other ideas. It was manoevred to the topmost set of falls and let go. But it never made it to the bottom. Half-way down the wild card came into play, and the block got wedged among the rocky edges of a cascade. At this point David graciously conceded artistic purpose to the more haphazard whims of nature, and reimagined the block as a water-swept boulder. It had been released, not just from the tree in which it had grown, but from his control.

Over the next twenty years it edged gradually downstream. In 1979 David found it beached in a pool, and covered with ice and dead leaves. In 1994 a flood sent it hurtling briefly downstream, carrying off a gate and part of a fence, until it became jammed under a road bridge. More stallings, rescues and free runs followed until that day in 1997, when I saw it one storm away from the River Dwyryd and a clear run to the Irish Sea. Now it was off to be the subject of other people's stories – maybe a minor nuisance to small boat-owners, or a topic for gossip among dolphins, or a cryptic message-in-a-bottle on a Jamaican beach.

'You've got to be about', they say here, if you want to keep pace with things, get back in touch. And in that suspended time between Christmas and New Year I made my first real shot at 'being about' again, making my own journeys out of the trees and into the wet. I went out in the jeep with an OS map on the passenger seat, seeing where lanes went, trying short cuts, looking for ways in. This was nothing new. All my working life I'd been an after-lunch nomad, ranging about, collecting bits of landscape flotsam and shards of stories, sometimes on quite deliberate hunts, more often as haphazardly as spindrift. This time I edged west along the valley, tracing the long string of fens from Roydon in the east, through the big swamp at Redgrave, down to the remote sedge-beds at Market Weston. Some of the vocabulary of these landscapes began to seep in: stands of alder like fretwork, goldfinches in silhouette among the twigs, looping flights of woodpeckers, wind-blown reed-mace fluff, unnameable rustlings in the reeds, and everywhere that sliding, glistening, immanent hint of water, stalking you, reminding you that even in the still moments of winter the world was 'being about', hatching plots and surprises.

I went east, too, seized by a fancy to find a winter heliotrope (the name means sun-following) in flower on New Year's Day. Polly and I had found some flowering precociously on our first trip to the Broads together, and it had become a kind of totem flower. I picked some for my room for its pink tassels and Mediterranean scent of honey and vanilla, and felt smug that I'd been able to track it down in our own parish – until I discovered that, had I gone a few hundred yards down the lane in the opposite direction, I would have found it blanketing the verges like a tarpaulin.

And I went north to see David Cobham and Liza Goddard. David had directed my first television film, *The Unofficial Countryside*, and we'd become good friends – a friendship which somehow survived his marrying my old love Liza. Now, they were both Norfok raptor guardians, and they took me to see Sculthorpe Moor, a new community reserve in the Wensum Valley. I imagined it would be much like the Waveney fens, but 40 miles is a long way between wild islands, even in Norfolk. Sculthorpe seemed more ancient and echoing than any wet place I'd seen. Huge willows had collapsed like broken sheaves, arching over the peat and here and there taking root at their tips. Their branches were draped – upholstered almost – with moss and lichen and epiphytic ferns. In danker spots there were mounds of tussock sedge, iron-dark and fusty, which in Norfolk were once cut to make fireside seats and church kneelers.

We meandered down to the river, found otter footprints (all five toes to the front, and known up here, somewhat confusingly, as 'seals'). We saw posts where barn owls had rested, and a thin trail that wound into the reed-swamp and just vanished – as had the escaped heifer that made it. It was wonderfully strange, and I felt like a landlubber. But then I saw something *I* knew, a bit of dialect I'd picked up back in wood country. Alongside the main track were what seemed to be groups of individual alder trees. But they were close together for their size, tilted outwards like sheave in what, when you looked closely, were loose, hollow rings. I seen exactly the same thing in the Chilterns, where anci hornbeam coppice-stools had lost their centres and many of t shoots. What looked like individual trees were part of a s plant. I guessed that these alders were also old stools, and at 7 or 8 feet in diameter could well have been five hundred years old.

This would make them the most precious and irreplaceable things on the reserve, and worryingly vulnerable in a place dedicated to open fenniness. So I risked appearing like a meddler, and mentioned my theory to Nigel, the warden. He was an intense but gentle Norfolk man, the grandson of the last professional sedge-cutter on Barton Broad, and he had eyes that sometimes focused hundreds of yards behind you. He was also, I suspect, a bit of a shaman, the kind of person who could conjure birds out of the air. He was the first to dream the reserve, while he was having his sandwich lunch nearby. He'd been entranced by the sight of marsh harriers displaying over the reeds, and had the nous to go and find out who owned the place. He discovered the lease was becoming vacant that week, and within a few days the Hawk and Owl Trust had taken over.

Now, back in his family tradition, Nigel was reinstating sedge-cutting and coppicing on a hunter-gatherer scale, and that Sunday morning this one-time parish common was filling up with local volunteers. The air was thick with Norfolk dialect, like the cackle of geese. John, poacher-turned-photographer, had found the remains of a vast nest low in a tree, and glimpsed an equally vast bird of prey. 'I was looking at this thing, and thinking thas too early for a harrier. Then it turned towards me, and I could see its thick wings and breast like a turkey and blust that was a goshawk.'

This, and the thought of marsh harriers in the spring, went to my head. I wondered what it would take to get the harriers, those perfect distillations of wind and reed, back to the Upper Waveney, where they'd last bred in the early nineteenth century. Restoring the fens again, maybe, all the way from Diss to Market Weston? I thought of the Wildlands Project in America, and their echoing mantra 'Reconnect Restore Rewild'. Back in the 1980s a group of environmental activists and romantic ecologists had a vision of joining up America's wild places, restoring John Muir's 'bee pastures' in the hamburger ranches, putting native forest back between (and, one day, over) the freeways: 'We live for the day when gray wolf populations are continuous from New Mexico to Greenland; when vast unbroken forests and flowing plains again thrive and support pre-Columbian populations of plants and animals; when humans dwell with respect, harmony and affection for the land.' It was too provocative and perhaps too selfish a vision ever to happen like that. But it has set up a dream to inspire more realistic action, an extreme position that is already pulling middle-of-the-road practicality in its direction.

Action on this scale is unsettling to us on this side of the Atlantic, used to a modest, defensive, deferential approach to conservation. But with a touch of poetic justice, it is the Dutch, whose engineers pioneered the drainage of East Anglia four centuries ago, who have taken up the baton and begun to give back to the wild much of the land they so arduously claimed from the sea. Their restored wetlands – tens of thousands of acres in extent – aren't anything like our tidy nature reserves. They've become a vast wet savannah of lakes, mudflats, immense reed-beds and scrub, teeming with birds – harriers, sea-eagles, spoonbills, and, in winter, fifty thousand greylag geese. There's virtually no human intervention. The water table is allowed to fluctuate naturally, and the whole area has large herds of semi-wild horses and 'de-domesticated' cattle. Soon these reserves will be joined up with wooded areas inland in what it is hoped will become 'a major unified nature-culture preserve, with the fewest possible barriers for people and animals'.

What makes a habitat, a lair, a place to live? The idea of a 'nature-culture preserve' is strange to us, so used to seeing these concepts in opposition. Yet it seems to me not only a desirable kind of dwelling (and something which is true to our roots as a species), but pretty much where I've ended up myself. Any human home – even a temporary one, like my room – is a cultural construction. But I came to mine almost by reflex, with as little thought about what I was doing as a migrant marsh harrier returning to the fen – 'naturally', if you like. If I'd consciously had to plan and choose where I was going to go in what was, for me, the most momentous change in my life, I would never have made it. That dithering between equally desirable alternatives would have been quite paralysing, a sure route back into my state of immobilising anxiety.

The freedom to choose where to live, what kind of life to lead, who to be, is regarded as one of the great privileges of being human – or of being a well-off First World human. But I've often found it disorientating, and maybe a block to more spontaneous, organic changes. To find a way of 'fitting' seems to me no more likely to come from deliberate choice than from accepting a degree of drift, from tacking with events, going with the flow. I'd fetched up, as a fledgling, in a situation I'd never dreamed of, in the simplest possible habitation, in a lair that felt, symbolically, like the primeval shelters humans made in woodland clearings. But it worked. I grew up fast. I got out of the house. I was being about again.

3

Commonplaces

> The commons is a curious and elegant institution within which human beings once lived free political lives while weaving through natural systems. The commons is a level of organisation of human society that includes the non-human.
>
> Gary Snyder, *The Practice of the Wild*

DURING THE LONG-DRAWN-OUT fag end of winter, I sat in my room with the cats, browsing through maps, grasping for little visions and intimations of spring. The experience of *le temps* is always in the context of place, and I was musing not just about when spring might 'occur' but where. Back in the Chilterns, my rites of spring involved a set of precisely located indicators, a series of stations on my own seasonal beating of the bounds: the first celandine by the river in Mill Lane, on the very edge of Berkhamsted, in time for my birthday in February (I cheated one desperately late year, and opened one up with a sun-lamp); the first bluebell in my wood (*weeks* before anybody else's); the first settled swifts high up over the parish church. What would be the equivalent up here? I prowled through the barns and outhouses wondering where the swallows might nest – if they came back. There were at least half a dozen old nest sites scattered through the buildings, including one perched, with almost Ruskinian grace, above a Victorian rocking horse. Would there be a bluebell anywhere? Or a neighbourhood orchid on the Ling? Might nightingales sing within earshot of my room?

I was marking time. So were the cats. After the intrepid hunters they'd so speedily become in the autumn, they had, in the cold, evolved into beanbags. Now they were showing signs of restlessness. They prowled about my room, drinking water from vases and bedside glasses, napping, then sitting up and intervening as soon as I appeared to be doing anything interesting that didn't involve them. 'Reading about the Stone Age in Norfolk? *I* want to, too. In fact I want to *go* there.' They loved a large Ordnance Survey map above all things, or a clamber across the electric typewriter (though they never succeeded in typing their names, as my Chiltern cat Pip, in a moment of serendipitous dancing, very nearly did). They each had subtly different ways of soliciting affection. Blanco was almost pure leopard, butting his head against mine, and patting my face with sheathed (usually) claws. Lily, delicately and almost apologetically, touched and licked the fingers of the hand she wished me to stroke her with. Blackie, the tart, simply collapsed sideways on the bed, like a slapstick comedian.

Thoreau confesses in his journals that whenever he was torpid, and in need of regeneration, he walked invariably towards the south-west. 'The future lies that way to me,' he wrote, 'and the earth seems more unexhausted and richer on that side.' This is partly a dig at the Old World, and what he saw as its obsession with history and dying institutions. But he thought he saw portents and signs in nature of an inclination towards the west. That was the way the sun moved and, he believed, migrant peoples and animals. It was 'the general movement of the race'. He saw 'westing' as a kind of primal instinct.

I'm not sure I share Thoreau's belief in a transcendental New Frontier, but the pull of the south-west is pretty compelling in the midst of a long winter, offering you the chance to meet spring on its way up, so to speak. Polly and I gave in to the temptation, and headed down to Cornwall. It is, they say, a foreign country down there, but it was shivering in exactly the same temperatures we'd left behind in Norfolk, and was just as thoroughly bashed about by flood and gale. Yet you could feel the pulse starting up. The first red campions and three-cornered leeks were in flower, and here and there precocious buzzards were chancing display flights. It felt, in the late January sun and salt air, a bit like southern Spain in winter.

One morning we meandered up the River Fal from the estuary, where I'd come in the 1980s to watch the March equinoctial tide, stained white with china-clay, rise up over the primroses in the hanging oak woods. The flowers had looked as marooned as beached starfish. There were no floating blossoms that morning, but another reminder that the movement of the seasons is always accompanied by migration and displacement, and the redefining of territories. Almost every creek had its sentinel egret, so dazzlingly white that it could have been carved out of alabaster. Little egrets were once just occasional vagrants from mainland Europe. But in the 1980s they began staying for the summer along the English south coast, and since 1996 have been breeding there. They may be one of the few gifts of climate change.

We wandered on, following our noses, and found a small quaggy valley, full of snipe. In the bushes we spotted a petite, olive-coloured bird with a strong eye-stripe. I assumed that it must be an overwintering chiffchaff. Then it came closer, and as the eye-stripes became more darkly complicated, I began to think we were in for a serious warbler twitch. But at 6 feet there was no doubt, though neither of us had seen such a bird before. It was a firecrest, a rare winter visitor to these parts. Its head was like a carnival mask, red-hot orange down the middle, dark oriental stripes above and through its eye, and a little bronze dimple just below its cheek. Then it did something extraordinary – it began to dance, soft-shoeing along a branch, jettéing into the air in loops and fly-catching parabolas, its wing-beats just slow enough to be visible. There was nothing especially of the west about it, but it seemed some kind of prelude.

There was a long-postponed Cornish quest to make while I was down here. I wanted to visit my name village of Mabe (and claim my inheritance, the Lordship of the Manor, etc.). But it proved to be a disappointment, a rather suburban one-time mining village, whose church was locked. It used to be dedicated to St Mabe, but now appeared to be the shrine of some impostor called St Laudus. We got the key from the vicar (it was Sunday, and we found him marching down the street with a bottle of Chardonnay under one arm) and poked about inside. There was something that looked like a Green Man carved high up in the bell-tower, with foliate beard and hair. Polly was sceptical, but having been humiliatingly de-sanctified I was convinced. Outside, as if in confirmation, there was an ancient Celtic stone, inscribed with a cross. The guidebook admitted it was a prehistoric nature symbol, and that (improbably) 'it was too difficult to move'. Instead it was diplomatically Christianised. I wondered if the man who gave it permission to stay was St Mabe, probably, like so many Cornish 'saints', just a vagabond Celtic nature-priest, westing his way down the peninsula, looking for spring.

*

Back home, I hauled out the maps again, trying to visualise the bits of the local landscape I hadn't seen yet. This was an old habit. Ever since I first came to East Anglia I'd been a compulsive map-worm. During the winters, or whenever I couldn't get out, I would pore over Ordnance Survey sheets, plotting walks, imagining the shapes of unknown, unvisited tracts, and sometimes just taking a sheer impractical delight in the abstracted patterns on the map. I had favourite imaginary retreats: High Wrong Corner (what on earth had happened there?) slap in the middle of a desolate stretch of Breckland; the strings of villages in Norfolk's heartland – Sall, Corpusty, Guist, Fulmodeston – their names strung out like concrete poems along the by-roads. And for some reason I became captivated by the shape of the country between Occold and Thorndon in Suffolk, where a long, and quite untypically bunched, grid of contour lines was cut across by strange, hieroglyphic field-boundaries. It had a faintly exotic, Arabic look, and I imagined the south-facing hillsides stepped with cultivation terraces, growing grapes and cherries, and maybe even peaches, as they had at Iken on the coast in medieval times.

Now the house I was living in was actually *on* the map, and I was ten minutes from the Hanging Gardens of Occold. I found myself driving through that much imagined patch by chance one day. There were, of course, no terraces or vineyards, not even an incomer's shot at a Mediterranean villa. But at the top of the hill was a vast and grotesque collection of green dolls' houses, a Lego set writ large. Up by the barricaded entrance the sign read 'Huntingdon Life Science' – our modern mark of the beast. That's the trouble with old maps (mine had been 1968 vintage): their freezings of time are manna to nostalgic fantasies.

'Out of the Map, into the Territory', David Abram urged in one of his provocative essays on the fuzzy edges between time and place. But you need somewhere to start. The prospect on the local map was as different from the Chilterns as it was possible to be. The contour lines were nonchalant. The lanes didn't so much snake as saunter. As for woods, they were chiefly wetland new-growth or plantations. Through half-shut eyes, the impression wasn't of the usual chequer of greens and browns, but of a lacework of blue, hatched with yellow. The line of the valley was clear. The Waveney wound back through the Ling to its source in Redgrave Fen, where the Little Ouse also rose, and flowed in exactly the opposite direction – due west to Breckland. There were tributaries in the fringes, mostly flowing south, so that there was a tongue of scattered fens and marshes, up to half a mile wide in places, between Diss in the east and Market Weston in the south-west. In the earliest maps they form an almost unbroken corridor of swampy land 10 miles long. Now, the combination of man-made drainage dykes and straightened channels gave the map the look of a sheet of cracked ice. Yet the water still connected things. As in one of those mazes printed in kids' puzzle books, I could trace the paths the water took, linking spring-lines, streams and field ditches back to their motherlode in the river.

I tried superimposing my personal map of connections and associations on top of this. We all have these inner atlases, irrational and hopelessly out-of-scale charts of landmarks, bench-marks and reference points. Mine made a shape something like an early satellite – an orb based around the valley, with aerials sticking out at right angles. One went north to the coast, to friends and marshland redoubts I'd known for much of my life. Another – strictly a through-route – went to the Brecks. The eastward prong ended up on the Suffolk coast between Aldeburgh and Southwold, where I had a cottage retreat for a few years, and joined up, through a skein of familiar villages, with the Stour Valley.

But this was just the framework, my mental arterial road system. In between was a latticework of shrines and talismans: particular dips in the road, tunnels of shadow, farmhouses invisible behind trees, a hamlet where every garden was full of crown imperials. And the more I looked at the symmetry of this pattern, the more it occurred to me that I might be living in the very centre of East Anglia. When I tried the idea out with a piece of string, I discovered that I was. If I'd cut out the shapes of Norfolk and

Suffolk and stuck a pin through the house, the map would have floated free, like a gyroscope.

Something else began to strike me, too. Much of the landscape seemed to be aligned along a north-west/south-east axis. The lean was most conspicuous in the east, where, in a stretch of country loosely centred on Diss and maybe a hundred square miles in extent, the grain of the old landscape – trackways, field boundaries, even the edges of woods – was unmistakably tilted towards the west. The direction was, in one way, neither here nor there. It was the synchronisation of the slants that was so intriguing, and the way that, in most places, they seemed to be entirely unconnected with topography or water-courses. The small fields south of one village, for example, were all aligned, but cut across the contours at every possible angle. Green lanes followed the general inclination even when it seemed to take them nowhere. Of course there were exceptions – close to the bigger rivers, or where the slopes were steep enough to command attention. The main Roman road to Norwich (now the A140) cut across the grain at an angle of about 25 degrees. But most other patches where the lean was in a different direction were obviously more modern features – the rectangular nineteenth-century plantations of some of the big estates for instance. Then I found a local historian who had measured the average inclination. It gave me gooseflesh: my new home range had a 4-degree tilt to the west.

The next day I bought a compass and went out to see if I could spot it, forgetting stupidly that the laws of perspective mean that you can never glimpse such a thing from ground level. Which made the whole phenomenon even more puzzling. If there were no sight cues how could it have been done deliberately? It strains credulity to think that the prehistoric farmers who first nibbled out the framework of this landscape should bother to harmonise their work. So was some fundamental fact of geography at the root of it, maybe one of the periodic slight falls or rises in the level of East Anglia relative to the sea? Or perhaps the prevailing east wind, or the transit of the sun, which made field-workers all 'tack' one way? Were they, in prehistoric times, more sensitive than we are to the earth's magnetic field? The lie of the land here is approximately directed towards the Magnetic North Pole. On the wilder shores of supposition, might the lanes have been pointing towards a major religious site, maybe Holme Wood Henge on the north coast, with the subsequent field clearances following their lead? Or was Thoreau right in believing that 'westing' is a fundamental urge in living things?

I doubt that there is a single reason for a grain in the landscape that may have been established in prehistoric times, when there were still wild horses and bison grazing in the valley fens. But I warm to the idea that this bit of rural planning wasn't done by landowners and bureaucrats on a drawing-board, but by some instinct, some unaccountable itch to turn towards the sun.

*

I'd first seen the horses back in the early autumn, on a day when Redgrave Fen looked almost russet under the low sun. I'd been watching two roe deer pick their way through the distant pools, when half a dozen heads jerked upright from the undergrowth, peered around like periscopes, and then vanished again. They had a dusty, taut, rangy look about them. It wasn't that I was surprised to see them. They were trumpeted on the reserve notice-board, mainly to discourage people from feeding them. What I wasn't prepared for was the jolt they gave to my senses, or the image they stirred of some other place and time. When, some days later, I saw them stream off at a gallop, reeds rippling round their shoulders, they looked like a herd of impala in the savannah. In fact Redgrave suddenly seemed like savannah itself. They had enchanted the place.

I've always kept horses at arm's length. I've never ridden, and never felt close to them. Impassive in the corners of tiny paddocks, barely shifting in hot or wet weather except to seek shelter behind each other's rumps, I find them sometimes too sad to watch. And my two abiding memories of horse encounters both have them in victim's roles. At a gymkhana in a Suffolk village in the 1970s, I'd watched a small child trying to take a low-jump course on her overwhelmed Shetland pony. They'd both failed miserably. The child's mother happened to be sitting next to me, and as the final fence fell, bellowed at her daughter, 'Take him round again. Don't let him get away with it!' That scene has hung in my mind ever since, a cameo of a whole pyramid of authoritarianism. A couple of years later, I watched my heavily pregnant and overdue sister-in-law Rose, stacked up with a chicken vindaloo, ride bareback round a paddock. It worked. Hannah was born the next day, the first day of spring. But that beast was a workhorse too, and the image of horses as the supreme example of wildness domesticated to the point of servility has stuck.

But the Redgrave horses didn't seem like that at all. They were a gipsyish, upstart bunch and I wanted to know more about them. They were introduced to the fen, of course, and to England for that matter. They were known as Koniks, and were brought here by the local wildlife trust to help keep the fen vegetation in check. Konik horses have a Byzantine history. They are descendants of Tarpans, the true wild horses of Europe, which clung on in the forests of Poland until the late nineteenth century. The species became extinct in 1876, when the last wild individual ran heroically over a cliff to avoid capture. But the genes of the Tarpan survived in hybrids which peasants had bred from them, and seemed to be vigorous and dominant. Quite spontaneously a strain of Polish farm-horses began to emerge that showed Tarpan characteristics: the mouse-grey ('grulla') coats, the black dorsal stripe, the dark manes that are almost Mohican in places, and which blow or fall naturally to one side, showing the blonde underhairs. In the 1930s these horses were the subject of dubious selective breeding programmes in Germany, in an attempt to re-create a pure 'Aryan' horse. The result was the Konik, to all intents and purposes a reconstituted Tarpan. With such a dark and complex history, touched with eugenics and echoes of antediluvian landscapes, it's no wonder the animals arouse strong emotions.

For conservationists the Koniks are chiefly just a hardy management tool. The local wildlife trust could have chosen upland sheep, or cattle. But the Koniks have marked advantages. They will eat anything – brambles, young birch, old rushes. They can walk anywhere, through dense reed and waist-deep water and cloying mud. They're able to help keep the encroaching woodland in check, and maintain a mosaic of open water, swamp and carr. They are virtually weather-proof (though they are given hay in cold weather), and they foal without help out on the fen. And as a bonus they have some kind of pedigree as a wild breed.

But some locals regard them, and the heavily managed fen which they help to maintain, as an alien and unnatural intrusion into a stretch of honest English commonland. Some even feel that the reserve is turning into a kind of wildlife park. Its star species – the fen raft-spider, which lives elusively in small flooded peat-pits – is given minute and bespoke attention. But that is a different order of intervention from laying on a herd of a half-domesticated mammal that hasn't occurred in the wild here for probably five thousand years. And the horses don't always behave as spirits of the wild. One morning, rather surprising ourselves, Polly and I found ourselves stalking them, like Apache scouts. Their almost circular hoofprints and even fresher piles of dung all led one way, and we found them browsing among the alders. There were seventeen of them (three geldings had formed a splinter group, separate from the main herd), and they were feeding in dense fen vegetation.

In a French magazine I'd found a reassuring description of their temperament:

> The Tarpan has a very calm disposition. They are friendly, curious and affectionate. The Tarpan is extremely intelligent. They are independent and quite stubborn. The Tarpan, unlike the domesticated modern horse, has not surrendered its freedom to man in trade for food and care, and therefore tends to rely more on its own view of the situation rather than allow its owner to make decisions for it. They seem to enjoy being ridden but they do not like being told where to go.

They were gipsies close to, as well as in the distance – quite small and with their woolly winter coats variegated in all kinds of pastel shades from grey to tan. Their faces were long and slightly donnish. They sidled up close to us and nodded extravagantly. We stayed stock-still. They sniffed us all over and washed Polly's cowering dog. Then one grabbed the zip of my coat and tried to open it. It could have been their 'curious and affectionate' natures but more likely a consequence of carrots in some earlier walker's pocket.

The whole experience here is full of that kind of illusion. The horses are free-ranging, but are also an enclosed matriarchy of mares and geldings, fenced-in and serviced occasionally by a peripatetic stallion. The fen itself survives in the midst of East Anglia's arable prairies only because of the equivalent of intensive care. Up to a few years ago it was going the way of almost all wet commonland. The lowering of the water table, aggravated locally by massive extraction for agriculture, was causing the fen to dry out. Birch and alder wood was colonising newly solid ground, causing it to dry even further. This of course is a process that happens quite naturally (though usually more slowly), and its end-result was to a lot of locals' taste. They liked the mix of dense scrub and swamp, the sense of wild encroachment. But it was ceasing to be the fen it had been since the last Ice Age, and in the late 1990s, its evolution was halted by a large grant from the European Community. The money was used to stop off the boreholes, clear the bush, scrape off the top layers of dried-out peat, and reflood the place. As a result it's become – if you ignore the surfeit of straight paths and interpretation boards and the absence of anything more predatory than a few conservationists – a passable facsimile of a prehistoric wetland. This is what the valley may have looked like in the Stone Age, when woods were still in their infancy in Britain.

And that thought broke my feeling that I'd seen these horses before. The Koniks are, almost to a T, the wild horses in Stone Age cave paintings. They have longer manes and bigger feet (the result maybe of generations of breeding for farm work) but their low-slung heads and pot bellies are instantly recognisable. The extraordinary Horse Panel in the Chauvet cave in the Ardèche, thirty thousand years old, could have been painted from life in the fen. Horses are far and away the most popular subject in palaeolithic cave art, and are usually wonderfully observed, sometimes in just a few lines, or by a clever use of the rock surface. The 'falling' or tumbling Tarpan in Lascaux is patched-out in bistre and charcoal right round the curves of a fissure, so that it appears to be rolling on its back, in the way that horses do, as you move round it. The 'Spotted Horse' panel in Peche Merle in the Lot has the muzzle of its right-hand animal drawn to echo a horse-head-shaped projection in the rock. The whole frieze is almost *pointilliste* in style, and was probably made by the blowing or spitting of a mixture of ochre pigment and charcoal, through splayed fingers. These animals are categorically not workhorses.

Ever since they were first discovered in the late nineteenth century, European cave paintings have been the subject of spirited and occasionally wild argument. Their locations, deep in dark and remote recesses, their purposes (if it makes sense to try and second-guess these), and their flashes of humour and caricature, all pose problems beyond the sheer wonder of their accomplishment. The Victorians, smarting at the humiliation this sophisticated art seemed to deal out to the presumptions of civilisation, wrote them off as meaningless doodles, the accidentally clever work of a few gifted copyists. Then, as the pictures' complex structures and techniques have been slowly unravelled, they've been interpreted rather like ink-blot tests, in ways that reflect the viewers' own preoccupations and prejudices. At the start of the twentieth century, anthropologists and ethnographers with late-colonial and utilitarian models of 'primitive' cultures had them down as purely functional hunting pictures, narratives of the chase or mammalian field guides. Or perhaps as magical aids to hunting, an envisioning and therefore 'capturing' (a word we still use of likenesses) of prey. Tarpans certainly were hunted for meat, often being driven over cliffs, as the last of their species eerily was. But they were also taken into semi-captivity for their milk, a precursor of domestication that was probably managed by women.

More recently the paintings were seen as fertility charms, intended to conjure up an abundance, not just of game but of all mating and growing things. Spears and genitalia were spotted everywhere, and the pictures began to be interpreted as palaeolithic pin-ups, male celebrations of sex and violence. Then the interest in altered states of consciousness encouraged a rise of shamanistic explanations, and the notion that the pictures formed a part of drug-, or dance-, driven rituals. More recently structuralists have begun to look at the layout and distribution of the art in a cave, not just at the individual portraits, and have found what seems to be a widespread association: bison (a male symbol?) are frequently painted on convex surfaces, facing horses (female?) on concave. They've suggested that the tableaux may be complex representations of the cosmology and philosophy of early Stone Age people.

Some of the theories were clearly not immune to biased selection and wishful thinking, and began to look decidedly shaky once the paintings were properly measured, zoologically identified, and, crucially, *traced* in their every scratch and marking, rather than impressionistically copied. When this was done some of the imagined spears, for instance, turned out to be bear-claw scratch-marks underneath the painting. Hunting magic seems increasingly only a partial explanation. There is, for instance, often an inverse relationship between the abundance of animals in a region, the likely staple food species (both deduced from fossils and the remains of meals in the caves) and the species which are most frequently painted. As Claude Lévi-Strauss says, totemic animals were not so much good to eat as 'good to think'. Carnivores, big cats and bears especially, are portrayed every bit as sympathetically as food species, and sometimes have the deepest recesses of a cave devoted to them – the earliest examples of an awe of great beasts that would stretch right through human culture. As for those feverishly imagined vulvas and erections, some have turned out, on dispassionate scrutiny, to be horses' hoof-prints and rather small fishes. There's not much doubt that their primary purpose was, in the broadest sense, religious. The choice of deep caves as galleries for the paintings, the use of the rock walls as a kind of portal to another world, the elusive sense of pattern in the arrangement of the paintings – all suggest some attempt to reach through to the animals' spirits, to the essence of nature and therefore the essence of life itself. Yet this is irrelevant in considering the quality and texture of the paintings. However serious and spiritual their ultimate purpose, their makers were working as gifted artists have always worked, creating art which was simultaneously affectionate, social, mythic, descriptive and plain playful.

I'm looking at a close-up photograph of the Horse Panel from the Chauvet cave now, trying to see with as little prejudice as possible just what it depicts. The four horses are on the right of a frieze of other animals, most of which, like them, are facing left. On one side is a group of aurochs, big-horned wild oxen, and beneath them two rhinoceroses squaring up to one another. (It's the only such encounter between these animals so far discovered in European cave art, and is made doubly remarkable by the way in which the torso of the right-hand rhinoceros is painted with deliberate distortion over a curve in the rock, and seems to swell or tense as it's viewed from different angles.) The horses are drawn in echelon, partly over these other animals and partly over each other. Their heads are tucked in close, as if they were in a photo-finish, and they look as if they may be a group of variations on a theme, different shots at catching a horse's head. Except that they are portraits, or imaginings, of four different animals. The top, and probably first-drawn, horse has its long head stretched out in classic Tarpan pose, with the facial muscles picked out and shaded by seams in the rock, and emphasised by scrapings with a flint burrin. The bottom horse is quite different, a Shetland pony by comparison. It is small, intense, its skin darkened with a mixture of clay and charcoal, and with its top lip slightly rolled back, in what seems an expression of surprise or curiosity (what is called 'flaagh' in horse-jargon).

Whatever the social or religious purposes of this painting it shares two incontestable qualities: a fascination with the animals, and an absorption in the protean business of painting itself. I don't know how one could draw a line between the two, separating creature empathy from creature depiction. A relishing of movement, individuality, nuances of mood, of the interplay between composure and spontaneity, seems to permeate both the forms of the animals, and the techniques with which they were portrayed. The first making of an image outside the head (however theoretical the idea of such an abrupt step is) was the first moment of culture, the decisive act which established the identity of the human species. Yet moments of gratuitous pleasure and reflection and rough conceptualisation must have occurred before this, and, later, at every stage between the preparation for the hunt and the finishing of the painting. Between, perhaps, the acute observation of the animals, the memorising of behaviour patterns and landscape mnemonics, and the quiet moments, the play of simple horse-watching and mimicry; and later, between the intimacy of milking and the first 'naming' of individuals. Even the premeditated ritual of painting must have been infused with the same sympathy of movement and feeling as animal-watching. The memory of a particular muzzle, perhaps, making a painter's face pucker up with pleasure; the inspired discovery that spitting could be a way of applying paint; then, from the serendipitous flickering of the animal-fat (horse-fat?) lamps used to light the caves, a half-drawn horse seeming to ripple into movement round a curve in the rock – the moment, as Annie Dillard wrote, when 'imagination can meet memory in the dark'.

On the evening of the winter's last frost, I saw seventeen wild horses galloping through the reeds, with their manes glowing crystal-red in the sunset. It seemed like a vision of East Anglia emerging from the Ice Age, a kind of cave painting itself. It doesn't matter a jot why these horses have been brought into this place, or whether this artfully stage-managed landscape has been arranged for utilitarian, scientific or romantic reasons. The horses have liberated it. Somehow, they've conjured up a wilderness.

*

There are no caves in East Anglia, nor even so much as a daub on a cliff. Our own landscape theatres — rather suitably, given the region's tradition of restrained ingenuity and general burrowing — are Stone Age flint mines. Grime's Graves is a cluster of nearly four hundred holes and pits in the Breckland, 20 miles to the west. Five thousand years ago flints were dug here on an almost industrial scale, yet the sandy plateau has something of the look of a deserted beach, pocked with humps and hollows that have been smoothed out by the tide. The poet Norman Nicholson, whose grandfather quarried iron-ore in Cumberland, once likened mining to the harvesting of a root crop that unfortunately didn't renew itself. Now other crops sometimes flourish between the pits: the miniature, fugitive spring flowers that are the natural covering of the old sandy wastes.

I've come here on a sunny afternoon in late winter. There are lapwings sweeing overhead, and fighter-bombers out of Lakenheath air-base, a reminder that back in the civilised world a war is brewing. Underneath me are the catacombs in the chalk where the locals hacked out the raw materials for their barbarous lifestyle: arm-ornaments, axes, borers, scrapers, burrins, harpoons, sickles, fireflints, sling-stones, tangs and barbs. It's bright enough for me to be allowed down into the main pit, a shaft 9 feet wide and 30 feet deep. All the way down the ladder the chalk walling is as soft as soap, and when I run my fingernail tentatively across it, I leave an incriminating scratch. So, I realise, have a hundred years of other nails. Somewhere among this scrabble of lines are the marks made by fire-hardened antler picks fifty centuries ago. There are two bands of flint in the chalk. Half-way down is a lode of dark, split stones a foot deep, shining, as if they have just been splashed with water. At the bottom is the floorstone, which held the most prized flints. There are little chambers and galleries branching off the main shaft, barred-off but artificially lit. The ceilings are very low. The diggers (were there children and women among them?) must have had to lie on their backs to prise out the stones. This far in they were away from the light which filters down the shaft, and used oil-lamps made of bone or hollowed chalk, similar to those found in the rock-art caves of southern Europe.

But why did they bother, when there was so much free-standing flint on the surface? The stone in the underground seams was certainly more amenable to splitting and knapping. Was it also more pleasing in its mineral shadings and mint surfaces? And were these hard-won stones, heaved out of underground chambers, smelling of gunpowder and metal long before either was discovered, seen to have more power, more *mana*, for hunting and making? In some of the chambers there are signs that something more social and complex went on here than utilitarian extraction: graffiti, carved chalk objects, piles of 'placed' funerary deposits. In one of the galleries, early explorers found the skull of a phalarope – now a rare migrant wading-bird in East Anglia – lying between two antler picks, arranged with their tines facing inwards.

Life finds its way into these deep recesses, one way or another. I'm pressing as close as I can against the barrier to one of the illuminated chambers, and can see that around the lights there is a growing smear of algae over the chalk. What will it be in twenty years, with the right breeze and a few more emissaries from the surface world? I try to imagine the sparse, brilliant, ephemeral plants of the Breckland sands, the speedwells and pinks, eluding the curators and taking root in the algal compost; a whole subterranean ecosystem maybe, like the steppe-land that still survived in the Brecks in neolithic times, when phalaropes may have bred among the heather and the pingos.

*

Flint is one of the most widely used raw materials on earth, yet no one is sure how it is formed, except that it is a kind of silica, leached out of chalk, and probably metamorphosed in the heat and pressure of volcanos. Back in the Chilterns, where countless stones seemed to rise spontaneously to the surface of the fields, they used to be thought of as simply sun-baked chalk lumps, natural rock-cakes. There is no more certainty about the date when the first tentative chippings were made into a stone that has naturally sharp, fractured edges anyway. I once found an eolith, one of those cryptic 'dawnstones' whose sharp edges may be signs of those first shapings, or then again be entirely natural, a product of pack-ice jostling – geology posing as craft. It was like the last two joints of a finger, with a point at one end. On the shaft something had flaked away an indentation like the mouth-piece of a flute, and on its journey to the surface this hollow had been filled by chalk in the perfect shape of a heart. It's my heartstone. I found it lying in my wood, close to where a ninety-million-year-old sea-urchin fossil, like a stone penny-bun, had materialised in the stump of a young ash tree.

The experience of history in the landscape is never in the tidy and purposeful order reassembled by historians. Layers are out of sequence, displaced, often flaunting a significance they haven't really earned. The valley's version of the sea-urchin in the tree is the raft-spider lurking in the flooded peat-pit. Or the communal washing-line on the Ling, strung together over the heather out of dead trees and recycled World War II concrete fence posts. Landscape, as a language, is pure pidgin. It's full of slang, neologisms, mimicry, and faddish jargon, yet with the odd knack of being comprehensible.

When I was 'in therapy' I once tried to outsmart my psychiatrist by likening the layers of stored memory to peat, a benign but essentially moribund medium. Memories, I argued, were like peat in that they were *over*. The echoes of disastrous relationships, corrosive habits, bad hard-wiring – surely, in a rational and self-aware person, these could float harmlessly in the mind, like fossil pollen particles caught in half-rotted reed-swamp? They could even be excavated for inspection, and debunking if necessary. But they were no longer active, part of the living, interacting surface of things.

It was a transparent piece of denial and a silly analogy. In the fens even the peat itself seems to be hard-wired, a persistent, nagging strain in the valley's life. One evening, free-range reading, I discovered that Virginia Woolf had spent a summer here in 1904, when she was twenty-four years old. She was staying at Blo' Norton Hall, and rode to Diss on her bicycle. She must have passed by our farmhouse. In her journal she described a watery landscape, humming with dragonflies and the marzipan smell of meadowsweet, and confessed to falling into the river ('though a walk in the fen has a singular charm, it is not to be undertaken as a way of getting to places'). 'It would need', she wrote, 'a careful and skilful brush to give a picture of this strange, grey-green, undulating, dreaming, philosophising and remembering land.' It's an arresting vision: Virginia Woolf as deep ecologist and premonitory river-dipper. But her labile imagination could hardly have failed to chime with this mercurial waterscape.

A thousand years earlier the anonymous namers of the valley settlements all paid homage to its pervasive wetness. Diss is from *disce*, Anglo-Saxon for 'ditch, or pool of standing water' (the Mere is still there). Redgrave is 'reed-ditch' or more likely 'red-grove' (alderwoods?). Hinderclay (*Hyldreclea, c.* 1095) is 'the tongue of land in a river fork where elder grew'. Thelnetham, 'the ham frequented by swans' (OE *elfetu*, for swan, with a common kind of syllable inversion). Blo', as in Norton, is probably simply 'bleak' or 'exposed', as in Clare's 'blea'. Back as far as the late Stone Age the valley seems always to have been a grey-green undulating mixture of swans and sallows. In the little hollows that were made by cottagers' peat-diggings, archaeologists have found the remains of willow-twigs going down through all the layers of peat, maybe covering ten thousand years.

Its formal history is scarcely more orderly. In the eleventh-century, Domesday records it as an area of few manors and many smallholders. Up until the early nineteenth century it was lived and worked in as more or less self-sufficient commonland. It was too wet to be anything else. Its inhabitants, like peasants across the planet, cobbled together a livelihood out of cutting peat for fuel, furze for bread ovens, reed and sedge for thatch, grazing a cow or a few geese, harvesting hedgerow fruit and a strip of bread-corn in the common-field. In the winter they were as amphibious as other fenland people, catching wildfowl and fishing for eels. The smallholders' chief 'export' was hemp, a do-it-yourself cottage-economy crop which was grown in tens of thousands of small plots and gardens. Many of the growers were also weavers, using their own home-grown fibres, which were 'retted' free of the stalks by soaking in the local ponds. South Lopham, on the Norfolk side of the big fen, made hemp linen by Royal Appointment in the nineteenth century. It was not a rich living but not one of wage-servility either. As J. M. Neeson put it, the commoners had 'a life as well as a living'.

The end of the economy (though not of the core of the landscape) began with the obliteration and cultivation of the local commons in the Enclosure of 1815–20. Much of the fenland was drained – though, acting out of either largesse or nervousness, landowners granted bigger 'Poor's Allotments' (which provided, as a kind of compensation, a few acres of ground for grazing or vegetables) than in many parts of England. And as capitalist economics began to take a grip in the newly rationalised landscape, so the versatile hemp faded out as a crop in the face of the more highly priced wheat.

What was left were little more than ritual echoes. In a sad parody of the complex customs of equitable commoning, the villagers of Market Weston were put through a little sedge-cutting game after Enclosure. A bell was rung, and the surviving commoners raced out to the fen to cut what sedge they could. A few hours later the bell was rung again to close the fen. Then the trustees moved in to mow the rest, distributing the cash it earned to the poor. The dole had arrived.

The scale of what was lost is all too clear on the earliest detailed map of Norfolk. William Faden, geographer to King George III, published his meticulously surveyed one-inch-to-the-mile map of Norfolk in 1797. It was the beginning of the most intense period of Parliamentary Enclosure, and the commons were given special attention, in the interests of ambitious landowners. In Breckland and the fenland near the Wash, they dominate the landscape. But there is scarely anywhere in the county where villages are not surrounded by common heaths, sedge-fens, wide drove-road edges (known locally as 'long meadows'), grazing greens and the last of the open fields. On the north bank of the Waveney a mosaic of commons, 4 miles wide in places, stretches all the way from Diss to Thetford. By 1850 they had virtually all disappeared under the plough or plantation forestry.

The landscape is full of echoes of this history, signposts to places that no longer exist. Almost every day I pass the drained-out arable deserts alongside Fen Lanes, the empty, cultivated spaces (or, these days, often uncompromisingly developed) of High Commons and Low Commons, the Lings where every sprout of heather has been exterminated. But that hard-wiring, that tilt towards the wild, can't be completely disconnected. The core of the valley is still a thin seam of inspiriting wildness, a 'Poor's Allotment' in a very non-materialist sense.

And then there are the moats, a real piece of surviving landscape dialect. We have moated halls, moated farmhouses, moated stackyards. There are piecemeal moats encircling entire village greens. Out in the fields, on the site of lost mansions, they lie like cryptic burial trenches. When they were first dug, they would have served a multitude of purposes. They were water reserves, cattle-troughs, flood-sumps, fishponds, boundary markers. Round individual houses they were doubtless status symbols, too, much like a modern gravel drive. Now hundreds of years on, they've become Water Features. Bungalows sport their own watery frontages, proud as pochards. Bits have been done up as ornamental ponds, with decking and cast-iron herons (and sometimes real herons). A few stretches have gone back to fen, among a froth of meadow-sweet and codlins-and-cream. The moats aren't any kind of inert fossil. They are a landscape narrative, as tenacious and adaptable as a good folk-tale.

The moat round Roger Deakin's farmhouse, one of the chain that surrounds the big green at Mellis, has one extra use. It's a swimming hole, circumnavigated by its owner most early mornings – and, on hot afternoons, by anyone else who can be persuaded in. Last summer there were shoals of lady writers bobbing among the duckweed and dragonflies. Roger is a true commoner, not just by dint of living on a common, but just, I think, by living. If it wasn't so pretentious I suspect he would put it down as his occupation. His life is suffused by the belief that, with ingenuity and a bit of respectful give-and-take towards fellow dwellers, anyone can do anything, in almost any company. Thirty years ago he rebuilt a derelict sixteenth-century farmhouse from scratch, learning every-thing from timber-hewing to plumbing as he went along. Now the farm is like a surreal arcadia: fields being teased back into meadowland, patches of new woodland (but many of the trees in rings not rows), barns full of corrugated iron and turned wood, found stones, the cockpit of a Canberra bomber, home-grown David Nash-style ash sculptures, graveyards of old Citroëns, vegetable patches and tree nurseries round the remains of an abandoned wood-mill ('the weeds help keep the roots moist'), converted shepherds' huts for sleeping in when the weather is too hot, and just in case the morning moat-crawl seems too forbidding, an outside bath. Roger is a believer in the layered life. Confront him with a group of Leylandii conifers, say, and his first instinct would not be to incinerate them, but to cover them with roses. He tacks round problems, almost always choosing the long, and therefore most eventful, way round. And out of this *chiaroscuro* life along Suffolk's damp seams, came his masterpiece *Waterlog*, a celebration of swimming as the true commonage of water, an intimate way into the second element.

Now he was working on a follow-up book on wood. (The wood and the water: how they haunt us both.) I'd barely seen him all winter. In the late autumn he'd travelled out to Kyrgyzstan, trekking in the mountains where the ancestors of our cultivated apples and walnuts grow. He'd stayed with the semi-nomadic tribes who camp out in the forests for the three months of the nut harvest, and live on a diet of wild fruit, honey, yoghurt and lamb. He'd come home with bags of walnuts bigger than his luggage, nuts in every possible variety of size and gnarledness, and a passable vocabulary of their local Kyrgyz names. Then he was off again, to the Australian rain-forests. The last I had heard from him was a call at midnight on New Year's Eve. He'd just been watching gum-trees exploding in the heat of Sydney's raging bush-fires – on the edge, as ever. I'd missed him while he was away. He is a source of perennial optimism and vision, and we talk in the same certifiable ecolect. 'Saw a hornet in the fen this morning.' 'Yeah, good hornet country here.' 'Probably its last outing. Hope the kids are around next year.' Roger's mischievous insect fantasy is to make an electronic organ based on the fact that crickets vary the pitch of their stridulations according to the temperatures. The crickets, encased in glass tubes, would be warmed and cooled by a keyboard control. In fact, of course, Roger wouldn't hurt a fly. When I was spending part of my rehabilitation time with him, a huge orange underwing moth blundered into the house while we were having supper. It couldn't be persuaded out, so we switched off all the inside lights, opened the door, and placed a large tilly-lamp out in the garden. We finished our meal in the glimmer, but the underwing flew to freedom, a fellow dusker.

Now Roger was home with a new, or rather revived, project. The saga of Cow Pasture Lane had re-surfaced. The lane is a droveway, part of an ancient network that links the local commons and greens. In the 1980s a farmer began to grub it up, doubtless believing his short-term occupation took precedence over centuries of common human business. Roger started a furious campaign against this impertinent stripping of the layers, and succeeded in saving most of the lane – though the farmer had earlier ploughed up another stretch. Now the local authority was taking an interest again, and thinking of upgrading the whole lane to the status of a byway, which would give them the legal grounds to reinstate the ploughed-up section. What it needed was more evidence of the lane's antiquity, its provenance, and Jane from the planning department was coming down to Mellis to hear our views.

So a few days later I'm down at Roger's farm for a reconnaissance and briefing meeting. His parlour is laid out like a war-room, with old maps and aerial photographs spread out over the tables. The lane shouts out from every one, as far back as 1783. But this is yesterday in old landscape terms, and we take Jane out to walk the track in search of more durable relics. And, desperate to impress, we start playing the scene like a three-hander from a Shakespearean comedy. Roger displays the topographical wonder of the lane, the sheer comeliness of it. I move upstage to woo Jane with bouquets of ancient flora. Roger smartly lays his historical cloak over a 'paved ford', which takes the track through a small stream. This has been a point of bafflement to both landowners and council planners, and Roger patiently explains: 'Not a pavement. Paving *stones*. Under the water.' We grub about and they aren't hard to find, three surviving flat boulders of a strange conglomerate rock that must have come from way beyond Suffolk.

But the lane hardly needs our Big Top presentation. It exudes pedigree and ecological style. All the way along it's broad enough for grazing, widening here and there into little bays where cattle would have paused for rest. The edges of the grass are full of the plants of ancient habitats: primroses, hart's-tongue fern, dog's mercury, even a single specimen of the hellebore that's known in Suffolk as setterwort (its root was introduced to the flesh of sick cattle by a thread or 'seton'). The hedges that line it aren't the planted strips of hawthorn you find in planned countryside, but the classic mixture of East Anglian old-growth woodland – ash, maple, hazel, with some hornbeam and oak. And they hadn't been laid, but coppiced, cut back to the ground level every eight to ten years as was the local tradition. Some of the stools, edging sideways over the centuries, are 8 feet across. As for the oak standards, there are barely two alike. There are crowns like lollipops and mushrooms. There are bent trunks, burred trunks, smooth barks, knotted barks, round acorns and bullet acorns – exactly the kind of diversity you'd expect from a wild, self-set population of oaks rather than a line of nursery-grown cuttings.

Cow Pasture Lane was an aboriginal droveway cut and trod out through the wildwood, which survived in marginal strips after the wood beyond had been cleared for agriculture. This was the common way that hedges were 'made' in the medieval period, but I'd hazard a guess that the Lane is a good deal older than this, probably Iron Age, like so much of the landscape's skeleton here. If so, it predates most of the local commons and greens (which may have begun as overnight resting-places) and is therefore older than the settlements which grew up around them.

Back home, map-bound and whimsical again, I tried to see how far I could extend Cow Pasture Lane using just green lanes and old by-roads. I tracked north-by-north-west (along the tilt!) up the Furzeway and Lizzie's Lane, to what was once Redgrave Common, through the fens, up the Broadway, linking the ghosts of Garboldisham and Harling heaths, into High Bridgham and onto the Drove, East Anglia's most ancient route-way, which passes through the one-time sheep-walks of Breckland all the way to Grimes Graves. One day I'll try and bike it. In the meantime, we learned a few weeks later that the council had agreed to designate the Lane.

*

That brief sortie into public affairs was a conscience-salver. I was all too aware that I was in danger of becoming unworldly, absorbed in the arcane business of the house and getting my own life in order. My scribblings still felt like a poor contribution to the world, and I itched to do something more assertive and conspicuous – make a cave painting, do up a shepherd's hut, rescue the fens. And through the lens of my little TV set the world beyond the valley seemed increasingly distant and insubstantial. Outside my window the rain continued to fall and the crop-sprayers made their umpteenth tour of the field, and I wondered, with the kind of self-satisfaction only available to someone who has learned to make bread in middle age, what on earth the misdemeanours of royal butlers and the erratic movements of the Dow Jones Index had to do with me. I was feeling close to the marrow of things. I was turning into a Little Fenlander. The words Jonathan Bate had uttered about the Romantics rang ominously in my ears: 'The price of this intoxication with the spirit of things is a definite break from the human community. Pantheism displaces philanthropy, communion with nature stands in for social awareness.'

But, to tell the truth, I didn't feel in the least bit smug, or socially detached. Less than six months after moving to East Anglia I felt back in touch, in control of my life again, grounded. I was making a living, busking bits of radio and journalism (I'd come back into print with an imperious review of two faunal camp-followers of *Flora Britannica*). And I was supporting myself emotionally to a degree I never had in my life before. Polly was my companion and comfort, but not yet my partner, and I had no choice but to be self-sufficient. And in an odd way I was feeling patriotic, not in some brainless nationalistic way, but from a growing fondness for my *patria*, my new home country. I don't think that a love of one's own place that bears no hostility to others is a bad emotion. It is a truly ecological one, and all living things are loyal to their own patch without being disrespectful of others. The valley and all its inhabitants were being generous to me, and I admired the maverick independence and creativity which had not only survived here, but seemed to be waxing strong.

So the old anarchist in me hunkered down with the new fenlander. And as Blair and Bush talked up their war against Iraq on the news, the business of politics and foreign affairs and 'reform environmentalism', seemed to drift into a self-referring circle, a perpetual repairing of the damage caused by the last ideological adventure. Through my porthole it looked uncannily like the political equivalent of iatrogenic disease, a chronic illness caused by treatments.

In the half-light of my oak-grove, I watched the 'nature' programmes for consolation. Blackie was miffed, and sat on my lap with her back to the television. Big cats are incomprehensible to their smaller cousins, who, firmly and practically rooted in their own world, will gaze intently, inches from the screen, when there are familiar sparrows or robins performing. I soon came round to her point of view. Sequences of carnivores pursuing game seemed like endlessly repeated film-loops, and caricatures of the complexities of life in the wild. Birds had miniature cameras strapped to their backs, so that we could 'share their view' of the world. Animated models of dinosaurs and cavemen, acting to scripts co-authored by Nietzsche and Barbara Cartland, sped through agonising family dramas towards their preordained destinies. Almost every programme, however honourably intended, seemed bent on belittling the natural world, on putting it firmly in its place. Animals were portrayed as playthings or bug-a-boos, or, alternatively, as mere bundles of sensations. There were series on *Weird Animals*, *Extreme Animals*, *Animal Killers*. The genre spreading most rapidly through all channels involved young male presenters, dressed up like Dr Livingstone, goading and tormenting hapless reptiles into retaliatory action. (Their dress sense was historically spot on; the Victorians acted in exactly the same spirit when they mounted their stuffed trophies with their fangs bared, just to show that they *deserved* to be killed.) Everywhere, cameras effected disorientating changes of focus and speed, regardless of appropriateness, so that pictures lost touch with authentic sense experiences. (I remember this being the basis for one of the few documentaries ever made about plants. It was called, predictably, *The Battle of the Leaves*, and used extravagant time-lapse photography to give the impression that wind-lashed leaves and twining stalks were literally fighting, a distortion every bit as anthropomorphic as the 'sentimental' view it implicitly derided.) And there, waiting in the wings, was the *reductio ad absurdum* of all this. ITV had produced a series called *Man vs Beast* that would have taken us straight back to the Roman circus. An elephant had been filmed competing against forty-four dwarves to pull a DC10, a bear and a man slavering it out in a hot-dog-eating race, and an orang-utang enduring a tug-of-war with a team of strongmen. For once, in the face of the tide of protest from the public, who are presumed to want such things, the series was put on hold.

I tuned in to David Attenborough's farewell saga, *The Life of Mammals*, hoping for better. But the carnivore loop was here, too, and the unquestioned assumption of a hierarchy of importance in nature. The series followed its subjects' increasing uprightness, climbing the Great Chain of Being towards the final programme and the triumphant Super Mammal, the one species that could make wars and TV documentaries. But in the penultimate programme there was a story that, at first sight, broke with the usual television stereotypes of aggression and competition. It was about a group of monkeys communicating and co-operating across the species barrier. Attenborough described how the different species in the different layers of the forest were attuned to spot the approach of danger, and give 'predator-specific' alarm calls that would send the whole mob scattering for their lives. He called it 'one of the most extraordinary anti-predator alliances in the world' (a trifle excessively; Blackie excites the same mass hubbub whenever she is out in the garden) and then dragged on a toy leopard, straight from Props'R'Us. The idea was to coax 'leopard-imminent cries' out of the monkeys. Fun-loving and sociable creatures that they are, they of course obliged, though for all I knew the noises may have been guffaws.

Not quite believing what I'd seen, I replayed the sequence. Attenborough presented it from the front, using that familiar, confiding whisper. He glanced knowingly over his shoulder. He was about to show us something remarkable, to draw back curtains. Then he pulled on the surrogate leopard. What the programme had conspired to produce was, not to put too fine a point on it, a freak-show, in the long tradition of menageries started by Phineas T. Barnum, which Barnum himself described, unaware of the resonance the name would have in the twenty-first century, as 'The Greatest Show on Earth'.

I recalled what the French writer Colette had said about these subtle exploitations of animals back in the 1930s. In a series of intimate essays she had been exploring her own relationships with animals. She had written about the sinuousness of a python, about listening to her dog's heart arrhythmias, about her transactions with a lizard her cat had brought in. Then, heartbroken by a zoo in Vincennes, she delivered a kind of credo:

> We are no longer free even to remain ignorant of how a panther, deliberately starved, rips open the throat of a goat, which, since the combat has to be spiced somehow and the cinema has no use for passive victims – has a kid to defend ... I shall dream, far from these wild creatures, that we could do without them, that we could leave them to live where they were born. We should forget their true shapes then, and our imaginations would flourish again.

I can't do without wild creatures, and suspect that our species can't either. To lose contact with our origins, with the wellsprings of life, with patterns of evolution and wisdom that are not controlled by us, with other ways of being against which to measure ourselves, with our *friends*, would have consequences we could scarcely bring ourselves to predict. To live with them only as dreams and legends is a vision of isolation that is hard to contemplate. Yet this doesn't deny Colette's argument that our relationship with wild creatures should be through imagination and respect, not through exploitation and manipulation and management – even if it is just stage-management. I wondered about the relationship the programme planners and makers imagined we and, more specifically, they had with the natural world. Were they stuck with being ringmasters, and we as being the audience, oohing and aahing in the wings? Unquestionably they've raised the profile of the natural world, stimulated interest in a general sort of way. But towards what kind of relationship? Is the mere purveyance of information the limits of the ambition of workers in this supposedly most expressive and flexible of media? Do they view the world much as the eighteenth-century makers of 'Cabinets of Curiosities' did, as a collection of diversions and amusements, to be attractively presented behind glass? Do they really believe that technological translations of the natural world – Slower! Closer! Bigger! – help us understand how we fit into it? Ironically, the aim of what have come to be called 'blue-chip' documentaries is to avoid any sense that human beings impinge on the natural world at all, or are even part of the same biosphere, despite the fact that the whole exercise of commercial film-making, its insistence that nature is all object not subject, its manipulation of story-lines, its knowing explaining-away of the complexities of behaviour, amounts to the most comprehensive impingement imaginable.

Four centuries ago, Francis Bacon, the writer who most clearly signposted the transition between an organic view of nature and the mechanistic and reductionist model of the modern world, wrote the script which the film-makers seem bent on following. Nature, he wrote, must be 'bound into service', be moulded by science. The 'searchers and spies' of nature are to discover her 'plots and secrets' ... 'Nature exhibits herself more clearly under the trials and vexations of art [i.e. science] than when left to herself.'

Since Bacon, the evolution of ideas about our place in nature has become a familiar and depressing story, and the 'presentation' of nature – something far from confined to television – can be seen as a sanitised version of a very old power game. If, historically, most ordinary people spent their lives in muddled mixtures of fear and wonder at nature, the philosphical tradition that stretched from Moses to Newton and beyond took it for granted that humankind was the supreme earthly order, and that the rest of creation had been put there for our benefit. During the Enlightenment of the seventeenth and eighteenth centuries, this assumption was expressed through an intense desire to understand and decode God's perfect design. Science, as Bacon dictated, would be the new medium of human assertion, its goal 'the enlarging of the bounds of human empire, to the effecting of all things possible'. Just occasionally, a rapt attention to nature's 'minute particulars' brought out, as it did in Gilbert White, a real respect for our fellow creatures. Yet most of those touched by the beauty and intricacy of the natural world still could not think of themselves as anything other than its focal point, the point of its existence.

During the nineteenth and twentieth centuries this fascination with the mechanics of nature led, inevitably, to a realisation of its interconnectedness and vulnerability – and that its intricate networks included *us*, whether we liked it or not. It was a revelation that should have stopped us in our tracks. It should have made us more tender, more cautious, more grateful to be open to the excitement and protean inventiveness of the planet. But the old feudal instincts were deeply ingrained in our culture, and soon re-emerged with a twist given by the imperial and paternalistic instincts of the Victorians. We would mature into the husbanders of nature. We'd look after it as if it were a farm – or a colony. Maybe, as the great biologist Lewis Thomas believes, this is the only realistic position:

> It is a despairing prospect. Here we are ... filled to exuberance with our new understanding of kinship to all the family of life, and here we are still nineteenth-century man, walking boot-shod over the open face of nature, subjugating and civilising it. And we cannot stop this controlling, unless we vanish under the hill ourselves ... We developed this way, we grew this way, we are this kind of species.

It was a depressing, defeatist conclusion, and even Thomas himself seemed uncomfortable with it. His 'solution' – long-term, large-scale – was to envisage a role for humankind as 'handyman for the earth ... someone to look after new symbiotic arrangements, store up information, do a certain amount of ornamenting ... That kind of thing.' (I once outlined Thomas's metaphor to Jim Lovelock, who gave it a neat Gaian tweak: 'Employed *by* the earth; yes, I like that!')

It's a more modest, useful role than landlord, true to some of our talents without closing the door to projects originating elsewhere in the ecosystem. But, for all Lovelock's benign sleight of hand, it's still essentially a mundane, utilitarian role, leaving in the air all our ancient and complex psychological ties with nature. As Bill McKibben has written, partly in response to Thomas's vision: 'That is our destiny? To be "caretakers" of a managed world, "custodians" of all life? For that job security we will trade the mystery of the natural world, the pungent mystery of our own lives and a world bursting with exuberant creation?'

The word most commonly used today to describe the role to which we should aspire is 'steward'. We are the planetary estate managers, with a duty to conserve and budget its natural resources. A steward is, of course, another party's agent, and just who we are stewards for is not always clear. Not often God, these days; more often humanity, our children, 'posterity', and only exceptionally nature itself. (Even this position isn't without its own brand of arrogance, when it's suggested for a biosphere that has managed very well by itself for epochs.) Mainstream environmentalism is unashamedly utilitarian and human-centred. It's based on enlightened self-interest: we want a healthy, unpolluted, species-rich ecosystem because our material future depends on it.

So of course do all organisms, and we've every right to our slice. The problem with stewardship is not the guaranteeing of our share, but the belief that we also have the right, or the duty, to determine every other species' share, too. The custodial relationship is intrinsically one of 'us' and 'them'. It assumes divisions, by power and importance, in a system that we know we must learn to see as a whole. However well-meaning, it allows back those authoritarian reflexes that are the root cause of the very ecological crises custodianship is trying to cure. Its inclination is to view nature as a static system, a collection of *things*, and to ignore dynamic *processes* – succession, habitat migration, natural catastrophe, nature's own recovery schemes and development plans. It favours caretaking above caring, and sees the comprehensive managing of the natural world as the goal of our 'naturework', not as a means towards a more equitable relationship. Above all, perhaps, it subtly reframes the object in need of custody as the non-human world, when perhaps it should be ourselves. The 'presentation' of nature fits snugly into this model, setting the non-human world on a plate, so to speak: fixed, packaged, knowable, consumable.

John Clare was not a presenter of nature. He was a re-presenter, a representative, a steward in a political sense. He wrote *out of* his 'special spots', rather than distantly, about them; and on behalf of his fellow creatures, never showing anything quite as facile as 'identification' so much as a kind of solidarity. At the very point where the act of writing might be thought to sever him from nature, he rejoins it in the special role of ecological minstrel, articulating and translating the songs of the marginalised majority. In 'The Lament of Swordy Well' he writes as if he were a 'piece of land', oppressed and exploited by 'vile enclosure' and 'gain's greedy hand'. In his long and mesmeric poem 'The Nightingale's Nest' he declares common cause with the bird nestling down in its workaday 'russet brown' plumage, and watches it sing 'where these crimping fern-leaves ramp among / The hazel's under-boughs . . .':

> . . . the happiest part
> Of summer's fame she shared – for so to me
> Did happy fancies shapen her employ . . .

Is Clare's relationship with the nightingale an example of a kind of cultural symbiosis? Their joint exultation reciprocated by his championing of the bird?

The best television documentaries emulate Clare's loving and diligent watchfulness. The patient vigils of cameramen exactly mirror his phrase: 'Hush, let the wood gate softly clap'. But it is hard to see in their dispassionate objectivity any hint of the poet's glimpse of a shared world, and of his sense of cherishing rather than caretaking.

Feeling jaundiced and gloomy I watched the news, and wondered for a moment if the programme had actually changed. The Prime Minister was presenting the case for a war against Iraq. His style was familiar: confiding, concerned, pious. He had a case, and would eventually pull back the curtains on the evidence. The Iraqis had 'weapons of mass destruction', the prerogative of school-prefect nations – though no one had succeeded in finding them yet. They were part of the international conspiracy of terrorism that he had pledged to crusade against. Trust him, he knew best, knew all. Any moment he would wheel on a model of Saddam Hussein for the British public to scream collective alarm calls against.

Perhaps watching the strutting of important men across the world's stage on television is not the most profound route to a view of human – and ecological – affairs. Perhaps, on the other hand, it puts them in a proper perspective. Either way, it was a pretty dispiriting evening. That Victorian image of Creation as a 'Great Chain of Being' was looking more and more like a Great Pyramid of Being, and felt as if it were getting pretty congested on the lower rungs.

Many of us thought long and hard about going on the great anti-war march scheduled for 15 February 2003. No one wanted to give comfort to Saddam, or to get back on the escalator of street violence that had alienated so many of us from political action in the 1970s. But there was a mood of palpable outrage in the country, a feeling that both decency and democracy were being shooed aside as inconveniences in the face of the Great Project. In the end two hundred of us went from the valley alone, joining two million others on the Thames Embankment in what turned out to be the biggest political protest in Britain's – and probably the West's – history. Nine-tenths of the protesters had never been on a demonstration before, and their mood was cautious and undog-matic, but resolutely determined – and, because demonstrations are also celebrations of shared beliefs and feelings, hugely festive. It was clear that what had brought this huge gathering of far from radical people onto the streets (the most popular poster – 'Make tea, not war' – caught the mood of the day exactly) was not just the war, but a growing sense of disenfranchisement. The premeditated violence that was being planned in our name seemed, in the banners of the myriad groups and factions, all of a piece with remote government think-tanks and unaccountable oil multinationals and the arms combines that had built up Iraq weaponry, and with all the apparatus of a ponderous, top-heavy state out of touch with its citizens.

I carried an extra torch in my heart, too, for the returning summer migrants – the swallows and turtle doves and cuckoos that take the eastern fly-way from Africa through the delta of the Tigris and Euphrates rivers, and up through central Iraq. They'd now have to contend with US and British bombers as well as Saddam's spray-planes along the thin chain they annually tried to forge between the poor south and the rich north.

*

The following month I went back to my old home ground in the Chilterns. I wasn't sure if it was a sensible thing to do. They say that you should always make a clean break, never look over your shoulder. But the Chilterns had been my life, my roots, the place that, for better or worse, had shaped me. They had been my exo-skeleton, my first romance, my raw material. I needed to go back if only to find out if their presence – that green wall up on the ridge – was still inside me, or whether it had been purged along with the less uplifting ghosts of my past. I thought that I was missing them, but wasn't sure. When I'd been ill my psychiatrist was always trying to persuade me to take myself up into the hills. He'd seen me once after I'd been roaming there for a while among the red kites, and glimpsed, I think, the person I'd once been, and what the birds had symbolised for me. He'd offered to take me, buy my lunch, bring me back. But even that seemed too much of a commitment. I shrank from the likelihood of feeling complete indifference, of sensing (to the extent that I sensed anything then) the total erasure of something that had once touched me so deeply. The high Chilterns and their free-wheeling kites were a touchstone. I had to dare to see if I had got them back.

The story of the Chiltern kites is a heart-warming one. The birds had once been common across England, before they were virtually wiped out by gamekeepers in the nineteenth century. They were persecuted for taking young pheasants and, bizarrely, for stealing washing to ornament their nests. (Autolycus in *The Winter's Tale*, himself 'a snapper-up of unconsidered trifles', warns that 'When the kite builds, look to your lesser linen'.) Only in the less barbarous redoubts of central Wales did the birds cling on, guarded as a symbol of wild Ceredigion. They made a slow but steady comeback, but expanding their range over Offa's Dyke looked a long way off. So in the late 1980s, the Nature Conservancy Council decided to give them a helping hand, and authorised the release of several pairs of Spanish-bred kites in the fastnesses of John Paul Getty's Chiltern estate near Ibstone – a pretty close approximation to the hilly savannahs they haunt in Europe. They seemed to like the place, stayed, and bred, and I saw my first (quite by accident) over the M40 in 1990. For a while they became a grail for me, an emblem of the wildest tract of country in the south-east, and sometimes, if I was feeling fanciful, of what I liked to think of as my Celtic ancestry.

So, that March afternoon I drove south on unfinished business, following, for a while, the roads I'd travelled every fortnight to see my psychiatrist. The hills were looking their late winter best – gaunt, reckless, poised for action. But it was a cool and blowy day, and I wondered if the birds would show themselves. I took my old route, a lane that runs along the bottom of the scarp, then, at the little village of Kingston Blount, turns abruptly east, up into the coombes.

And there they were. They could have been any big birds at a distance, drifting in the grey sky. Then they lifted up, flexed, soared, two taut crossbows against the leafless ridge-woods. They glided towards me – no hurry, just riding the wind, sliding across the eddies. They came close, and I could see the rufous plumage ruffling on their bodies and tails. They were calling, but the wind carried their cries away from me. I drove further south, up onto the plateau. There were kites everywhere. They were sporting over the villages, lifting on gusts that took them sailing clean over cottages then down to the level of the bird-tables. When I stopped for lunch in a pub, I could see them through the windows, arcing across the hedges, huge and buoyant. Outside I watched one close to as it turned into the wind. It raised its wings – as relaxed as a dancer's arms or a half-full jibsail – and gathered the air in, folded it into itself. It was so poised, so effortlessly muscular that I could feel my own shoulders flexing in sympathy.

The light was failing. I took a short-cut for home, up a lane through some rough pastureland. Suddenly the air was full of kites, a shifting mesh of flight-lines that stretched as far as I could see. It was, I guess, one of those excited gatherings that, for so many species, are part of the ritual of roosting. But that explains nothing. This was a wilful, gratuitous relishing of the wind. All birds of prey love the wind. It is when they become what they are truly made for, reach a kind of epiphany. Yet this awesome, communal display was more than that. These birds were daredevilling, taking their flight skills to the edge.

I was rooted to the spot. I couldn't count the birds. There were thirty, forty in front of me, and, when I turned round, just as many behind. They were exalted, falling out of the sky like peregrines, skimming the fields, stooping, spiralling, stalling, their forked tails fine-tuning their balance so effortlessly that it looked as if they were juggling the wind. Each time one rose or fell, I felt myself rise and fall with it. I was deep-breathing in sympathy. I didn't want to fly, or be up there with them, but they transported me back to being six years old again, rushing down a hill with my arms held out – and back, maybe, very much further than that.

What was going on? There were a few playful chases, but this was no mass courtship display. Nor was it one of the gatherings that now occur round the Chiltern villages where, with the kindest of intentions, householders are putting out steak on their bird-tables (and incidentally putting the birds' digestive systems in peril). This was tribal cement, a carnival of solidarity at the end of the day, a display of mutual confidence-boosting in the element they knew best, just as we might see the old year out in a tempest of dance.

The light faded, and the birds began to melt into the trees. I followed them, past Getty's extraordinary cricket pitch, which sits like an ornate island amidst the louring beech hangers. I heard the thin seeps of tit flocks shuttling to roost, and then, just above me, a kite's mew. A single bird, just a shadow now, was gliding through the branches. We broke cover together, and on the edge of a twilit clearing it swooped down like a smothering cloak; no longer a dancer but Accipiter, the one who takes to itself.

*

Going back to my wood was tougher by half, and I didn't feel in the least bit like dancing. Its sale was imminent, and the lurking prospect of change was causing painful waves. The village was uneasy. Big landowners, local and foreign, were rustling their chequebooks. It might become a hunting province again, or a paintball arena, or scoured out by big-hoofed trials horses. My friends Francesca Greenoak and John Kilpatrick, who'd been guardians, helpers and irreplaceable pillars of support since the beginning, were rapidly putting together a local trust to buy the wood, but were quietly furious at my negligence in having made no provision for an emergency like this. What I felt myself, I wasn't sure, and didn't dare think about it until I'd been back. The need to try and put my life back in order had given me a kind of tunnel vision.

The omens, that March weekend, weren't good. There was a 'For Sale' notice pinned to the big beech tree at the entrance – the same notice from the same agent on the same tree that had given me gooseflesh, like a prayer answered, twenty years before ('Wood For Sale'!). And there was a Ford Estate wrapped round the ash tree next to it. These joy-riders' relics used to slide off the s-bend in the lane most winters. But this one was different, portentous with the shreds of some half-finished story. It was a painter-and-decorator's van, full of equipment, and in the impact the paint pots had hurtled forward and erupted, coating the inside of the car with a pale viscous fluid, like a reptile's blood. It looked like a road-kill itself, pathetic and reproachful. 'Look what you did!' it said. 'You didn't look where you were going! You ran away!'

I had to remind myself that I'd never intended the wood to be a piece of personal property, a possession. I owned it, I once wrote, only to *dis*own it, to give it back to its rightful citizens. But, looking at the graphic clarity of that agent's sign (it was eerily like the notices that were posted round the village when the paths were stopped up at Enclosure in 1853), I remembered what it had been like the last time, when I was the prospective purchaser, not the vendor. I'd hidden among the trees, watching rival bidders touring the glades with their peaked caps and clip-boards, and agonised over the possibility that it might not be mine. Had I been disingenuous all those years, hiding an ego-trip under the cloak of ecological altruism? I'd wanted to set up the field of play for a community experiment, and see if a bunch of twentieth-century villagers could go 'back to the woods', make good the damage of commercial occupation, and enjoy themselves. I'd hoped too that it could be a kind of workplace, though I would never have had the brass of my friend Tony Evans when he suggested to the Inland Revenue that he might charge the entire cost of his wooded estate, his 'outdoor studio', as a legitimate business expense.

But had I been looking for a kind of studio too, a place to 'pose' both myself and the landscape? Or a retreat to make up for my failure to make a conventional dwelling of my own? The village, my friends, and a large extended community of frogs, sparrowhawks, badgers, orchids, monkjacks and nomadic bats had worked and played here energetically. We'd cleared out the inappropriate plantings of the last owner (matchwood poplars), let in the light, freed up regeneration and created a network of paths simply by treading them out, often along the tracks already made by animals. We'd experimented with some on-site democracy to decide where the glades should be and evolved a rough fuel-for-work system for the firewood we produced. But in confessional mood, I was uncomfortably aware how much of what we'd done had been initiated or moulded by me – when and where we worked, what were to be regarded as the significant patches of vegetation, the decision to cash-in the capital represented by two hundred timber beeches. I was, as so often, riding that tightrope between simple engagement and subtle control, only this time a lot of my own species were part of the constituency. And I'd made no provision for their future, for the stake they had in the place. Was I afraid of acknowledging my own mortality? Was the wood – an enduring, resourceful place – some kind of counter to that too?

But that March day the wood, as wild places often are, was generous. It let me off the hook. So did Fran and John. Their Trust was shaping up well. It was on the point of receiving a Heritage

Lottery grant, and poised to secure the wood's future 'in perpetuity'. And they'd seen red kites over the beeches, drifting up from their motherlode 20 miles to the south, a vision maybe of the future. The wood itself looked new-minted, rinsed clean by yesterday's gale. The bluebell shoots were already 6 inches high, only a few weeks away from flowering. Once, on my hands and knees, I'd found seventeen colour variants here, from pure white and white stripes on pastel blue, through to darkest indigo. Another afternoon, I'd surprised a pure white stag snoozing in their blue wash, as astonishing a sight as a unicorn. I'd watched children sleeping among the flowers too. The children were our keepers of the faith. They were indifferent to our attacks of work ethic, and held on to the best sort of unselfconscious animism. I'd seen them lost in deep conversation with trees, or earnestly escorting frogs back to their breeding pond — 'in case they couldn't get over the branches'. I looked at the tracks we'd made, the glinting, grey sheen on the beech-trunks, the sheaves of new saplings coming up in the thinned areas, and wondered if I was overstating the superior vitality of my new wetland home.

And I realised that the sense of having some kind of responsibility for the place had bred in me an intense attention to its evolution. Back in the days when I walked the tracks two or three times a week I could spot the most infinitesimal changes. I knew when individual ash seedlings had grown a single inch, where windfall branches had dropped from, exactly how big our colonies of wood vetch were, where flash-floods would travel, where the badgers had walked the night before. And now I found those two decades of ongoing memory suddenly visible again, as if I had drawn out a peat-core from my mind. The glades we'd cut out by clearing the last owner's plantation poplars were now soaring, naturally regenerated woodlets. Three years ago they weren't much taller than me. We scarcely planted a single tree, but close on ten species had come in of their own accord. And underneath there were already tufts of colonising spurge-laurel and shield-ferns. I walked on to the top of the wood. I'd often worked up here by myself, ring-barking the more objectionable poplars so that, in their fallow years, they could at least be useful to tits and woodpeckers. Now I could see how far the trees had collapsed, the way the young oaks beneath were insinuating themselves between the dead branches. It was, over maybe half an acre, just like a real forest.

But there were more playful touches, too. The mistletoe whose fruit I'd rescued from a doomed orchard in the Vale of Evesham in 1982, and rubbed into the bark of our one and only crab apple, was now all of 4 inches long. The ash seedling that we'd naively put a tree shelter round was a miserable one-quarter the size of its unprotected neighbours. And the oakling lawns that had sprouted after the great acorn autumn of 1994, promising such mighty trees, had all vanished – eaten by the caterpillars parachuting down from their parents. Woods have ancient and mischievous rhythms all of their own.

The one piece of serious landscaping we'd done looked as if it had been there for ever. We'd nervously cut a track into one of the wood's complicated slopes in order to provide vehicle access, wondering if it was a presumptuous imposition, a contradiction of what we were trying to achieve. We needn't have worried. It had rapidly evolved into a very social, or perhaps I should say common, landscape. I'd made the decision to create the track. The village children helped ease its impact by taking ferns and flowers home to foster for the winter. Our digger-driver added a slight but seductive curve that we could never have dreamed of, and which let in the sunset in the spring. And a larger throng still had mellowed the whole thing, treating it like any other minor geological disturbance, inhabiting it, untidying it up. First, the algae on the bare chalk, then the bluebells, then the billows of old-man's-beard and our treasured ticking-striped wood vetch – they simply absorbed our folly. Now they were carrying on with their own agenda. The bank had turned shady, moss-rimmed. The wood vetch had migrated 20 yards into the sun. But the track was still plainly a grace-note we'd added to the place, and I felt absurdly touched that even I – the captain who had jumped ship – had left a small trace.

But I was even more gratified that the Chilterns still had their ineradicable trace in me. Going back hadn't filled me with regrets at leaving, nor unleashed any uncomfortable memories of being cocooned or moribund. They seemed rinsed clear of any associations of cosseting or sickness or retreat. They were – and would always be – simply the place I knew best. The Chiltern ridges and coombes are often likened to a clenched fist, and their topography rescues the cliché: I know them, and especially Hardings Wood, like the back of my hand. And I know I can always go back, out of love now, not bondage.

*

It's dusk again. I'm trying to think of a joke about my 'common room', but it won't work. One's own room (what Gilbert White called his 'hybernaculum') is the least shared of all spaces. (This is true of the non-human world, too. Only under the stress of cold weather do most species abandon their desire for a smidgeon of private space when they're resting.) But at some indefinable distance beyond, the idea of space and land as private property softens: 'the Englishman's castle' backs onto the English people's birthright. These perfectly contrary views are held unashamedly across the planet, and a belief that land is a kind of common inheritance is deep-rooted, and maybe there instinctively from the beginning of life.

When I was a boy the commons began at the end of the garden, separated off by no more than a single filament of barbed wire. Our road of detached 'residences' had been built in the grounds of a demolished Hall, and the one-time landscaped grounds lay, unclaimed, between us and a new council estate. Its hundred or so acres were returning inexorably to the wild, and the exotic cedars and silver firs were under siege by a riotous tide of thorn and thick grass. It was a backyard savannah, and, as feral as all kids were in those days, we immediately annexed it as our natural, biological home We called it simply 'the Field', not from any sense of its primordialness, but because it was the only one we had or needed.

It was owned by the man who had built our houses, and once in a while he harrowed a bit of the rough grass or put out a few cattle. But he seemed happy that the entire neighbourhood, young and old, used it as their common. What I find intriguing is how, quite untutored, our gang evolved an equitable pattern of mapping and colonising this territory that meshed not only with his fits of hobby-farming but with our parents' more utilitarian uses (it was a short cut home from the station). No one presumed to be clan leader, or to claim any special privileges beyond those that came anyway from superior age or strength. But somehow we tricked out a geography for the place, a network of paths, sites for meetings and camps, no-go areas, ritual arenas. We devised a kind of peasant economy, too. We recycled bricks from the rubble of the old Hall, and foraged for walnuts and chestnuts – and almonds, too – from the ornamental trees along our road. We cooked potatoes over birch-wood fires, and learned how to make butter by cranking a jar of cream on the wheel of an upturned bicycle. We had an attentiveness to things that was close to animism: to the texture of promising firewood, to imagined faces in tree-trunks, to heroic flints, to the suppleness of grasses and all the tortures they could be used for, to the tastes of leaves, to dips in the land that were sufficient to hide us from our parents when we were 'up to something'. We stayed out there all day in the summer holidays, often coming home only for tea and sleep. As a place of habitation it had only one real restriction: it was tribal, our gang's territory and nobody else's. When buccaneers from the council estate strayed in (they had their own patch north of the Field) they were dealt with ritually, but summarily.

The town's real common, up in the hills, couldn't have seemed more different. To me it was a foreign and hostile place. It was where I was sent on long-distance runs when the weather was too wet for rugby, and I loathed it for its association with loneliness and drudgery and winter. But the town didn't share my childish distaste for the place. It was one of the few issues over which opinion was united across the classes. Berkhamsted Common had been liberated from a cynical and illegal Enclosure in 1866 by a spectacular piece of direct action, and the townspeople weren't about to forget it. But their legal rights were vestigial, even though the great tract of bracken and furze had been commonland since time immemorial. When the Brownlows (lords of the manor) sold up in the 1920s, half the common was bought by the National Trust, and the other by the local golf club. A few years later the club conceded rights of 'air and exercise' to the public under the provisions of the Metropolitan Commons Act. But the relationship between landowner and commoners remains an edgy one, as I described in my book *Home Country*:

> Local gleaners haunt the heather, looking for golf balls which aren't always entirely lost. Non-golfers out for strolls brave the curses and stray balls of the foursomes and stride across the fairways. It isn't so much their legal rights they are asserting as a sense of the common as a shared

inheritance, a sanctuary, a seam of 'good ground' standing watch over the town. I once heard a High Street butcher, on a stiflingly hot day, explaining to a customer that he always kept the back door of the shop open to 'let in the fresh air from the common'.

Later I learned to love the place, too, but more as an academic admirer than a commoner. (My own common would always and for ever be that ragged field from the 1950s.) I read about the human sensitivity and ecological wisdom of the customs by which the place operated, the close seasons when the cutting of bracken and gorse was forbidden, the regulations about the size of billhooks permitted, and the exceptions made for the over-sixties and under-fourteens. I read Vinogradoff's majestic *Villeinage in England*, with its analysis of how, in its heyday, commoning was a balanced, closed-energy system, with the fodder that was grazed off the common itself 'paid for' by the animals dunging in autumn on the strips of the common-field. I saw this anciently equitable system being argued about on modern Dartmoor, in a spirited debate about whether bees, the epitome of wild foragers, could be regarded as commonable animals. If Friends of the Earth's slogan is 'Act locally, think globally', the commoners' was 'Act locally, *think* locally'.

This may seem insular now, but it encouraged independence and frugality and a neighbourly intimacy with other commoners, of all species. The French historian Pierre Bourdieu has coined the word 'habitus' for this kind of shared dwelling: 'a lived environment comprised of practices, inherited expectations, rules which both determined limits to usages and disclosed possibilities, norms and sanctions both of law and neighbourhood pressures'.

*

This, I think, is how the Ling originally worked, back in East Anglia. I can almost see the edges of its 250-odd acres from my window. Lying in the original flood plain of the River Waveney, it was once covered with peat, like the local fens. But by the 1840s this had been dug out, exposing an expanse of almost pure sand. It was too infertile to be taken into cultivation, even during the Dig for Victory campaign of the last war, and in the 1940s and 50s was, paradoxically, both wild and used. Roy Potter and his family have lived on the Ling for generations. He's a master-builder in his day job, and plays a Stratocaster in a rock revival band in the evening, and has vivid memories of those post-war days. The heather and gorse were like thickets in places, and teeming with birds, skylarks especially. The men shot partridges and rabbits, the boys ran wild, and the women put out their washing on the communal line. Everyone dug up domestic quantities of sand and gravel when they were needed, cut any encroaching trees for firewood and occasionally burned parts to the ground.

But it was a far-from-workaday place. In hard winters the shallow ponds froze, and the village went skating. Roy still has his wooden skates hanging on the door. In summer, the local children went off in a gang to swim in the river, which – un-canalised then – still had sandy banks. Their mums used to summon them back at tea-time by blowing a whistle. Before the water table had been sucked down by extraction boreholes, the Ling had strange and hallucinatory mists, too, plates of vapour that hovered at head height, obscuring all the geography below. On the Ling on one such night, Roy's father, a little the worse for drink, arrived home on the wrong doorstep. Peering above the mist, he'd seen what looked like a familiar light in the distance, but none of the country in between.

The common, as all birthplaces are, is a stage and a lens. Down there, everything is sharper, more dramatic. The frosts are more devastating, the floods deeper, the snakes vaster, the unexploded bombs more devilishly hidden ... We, of course, josh Roy wickedly, and 'On the Ling...' is now our response to anyone, from weather forecaster to returning traveller, who has the gall to predict or claim exceptional circumstances. But he's right. The Ling's weatherbeaten surfaces have a kind of bravura: anything could happen there.

But the place has changed. The old common usages, both formal and informal, have been outlawed. No one can shoot rabbits or dig sand, or – except the new conservation managers – cut trees. The town of Diss, only 2 miles away, has expanded, and with it the pressure of car-borne vistors. And when the county wildlife trust took over the lease they subtly altered its character. Acting on the premise that heathland is a rare and special habitat, and needs bespoke management to maintain it, they cut the long grass and brought in brush-cutters to reduce drastically the areas of gorse and tall heather. The aim was to encourage the low-growing plants of sandy heaths. The Ling was going specialist.

Scrub is the enemy of official nature conservation. Despite being an entirely natural habitat, the haunt of nightingales, breeding warblers, roosting winter birds, shy orchids and a multitude of insects, its removal – or at least control – is the priority on almost all nature reserves. There are often good reasons for this. Scrub is simply burgeoning woodland, and is a stage that all habitats – heaths, fens, chalk downs – will eventually evolve to if they're not cut back by fire or grazing animals or human intervention. But there are times when it seems almost to have become demonised. Perhaps its very wildness, its ragged, unpredictable luxuriance – the result not of management but spontaneous growth – is what has blackened its image.

But on the Ling, its removal brought the wild card into play again. The place was rapidly invaded by legions of rabbits. Rabbits are usually welcomed on grassland reserves as small, self-supporting graziers, which keep ranker vegetation in check. But these came in huge numbers, attracted by the ease of grazing on the new short turf, and now patches of the Ling, cropped of most of their vegetation, are starting to erode. To be fair, there have been gains as well as losses. The plants of short turf – sheep's sorrel, bedstraws, stag's-horn lichens – have prospered. Dog walkers enjoy the easy access and open prospects. But the variety of plant and bird – and human – life has declined. It's a moot point whether the old system of common usage – the happenstance burning and tree-lopping – mightn't have been a better mode of occupation, for commoners of all species. When it is removed from a concern for all local organisms, even conservation can be a kind of monoculture.

The commons system was never perfect. Local populations grew, grazing pressures became more than the land could carry, rogue individuals – insiders as well as outsiders – would sometimes buck the laws and customs of responsible use. Normally they were dealt with by the manorial courts, or by the sanctions of neighbours, and the copious records that exist of these transgressions give an unfair picture of the problems of the commons: when they were working well it simply wasn't recorded.

What didn't happen was some internal collapse of the whole system. This is what was portrayed as the inevitable fate of 'un-owned' resources by the American ecologist Garrett Hardin in the 1960s, in his portentous and Calvinist but seminally influential essay 'The Tragedy of the Commons'. Hardin argued that any natural resource that isn't subject to the restraints afforded to private property will always, ultimately, be over-exploited, possibly to the point of extinction. 'Each man is locked into a system that compels him to increase his herd without limit – in a world that is limited. Ruin is the destination toward which all men rush, each pursuing his own best interest in a society that believes in the freedom of the commons. Freedom in a commons brings ruin to all.' His pessimism might have been justified in the case of the exploitation of 'virgin' resources by nomadic industrialists (for example, in logging and factory-trawling). But he clearly knew nothing of the commons system in England, or for that matter in peasant societies throughout the world, where rootedness and neighbourliness made self-regulation second nature. And he was using, interestingly, the same argument that was trotted out by apologists for the Enclosures in England. The taint of greed in man was ineradicable, and continuing to allow free access (especially to the poor, with their primitive absence of morals) to natural resources would bring economic calamity. What the enclosers meant covertly, of course, was economic calamity for *them*. But this kind of disaster was an ideological fiction. In almost all instances what caused the demise of the system was not its inefficiency but its appropriation by outside forces.

The reality of what was happening is enshrined in the legal framework that grew up to legitimise the seizure of the commons. The fundamental assumption was that common rights were 'granted' by some theoretical landowners in the past, and could therefore just as summarily be withdrawn. Just what state of affairs existed before that moment was kept deliberately vague; the landowners seemingly came into possession of their estates with the mysterious abruptness of Adam's inheritance of Eden. This legal myth made the law not an arbitrator, but an instrument of appropriation. 'The law pretended', E.P. Thompson wrote, 'that, somewhere in the year dot, the commons were granted by benevolent Saxon or Norman landowners, so that uses were less of right than by grace . . . it guarded against the danger that use-rights might be seen as inherent in the users.' It scarcely needs a historian to confirm their real origins, and that legalistic common 'rights' were simply the politically approved remnants of a much broader system of common usage.

Yet the legal machinations did even deeper damage than foreclosing ancient customs and usages. They made those that survived conditional on property-owning, not dwelling. The freehold cottager might have rights on the fen, but the tenant might not, and certainly not the landless poor. The traditional usages – acquired simply by the fact of living in an ecosystem – were translated into property rights. They were 'reified', turned into things, in a process that is exactly analogous to the way that nature itself has been turned into an object.

E. P. Thompson talks of a 'customary consciousness', in which rights were asserted as 'ours' rather than 'mine' or 'thine'. John Clare's customary consciousness included the non-human, and, in Thompson's words, he 'may be described, without hindsight, as a poet of ecological protest: he was not writing about man here and nature there, but lamenting a threatened equilibrium in which both were involved'. His Enclosure elegy 'Remembrances' (p. 38), continues with these lines on a keeper's gibbet:

– O I never call to mind
These pleasant names of places but I leave a sigh behind
While I see the little mouldywarps hang sweeing in the wind
On the only aged willow that in all the field remains
& nature hides her face while theyre sweeing in their chains
& in a silent murmuring complains.

Here was common for their hills where they seek for freedom still
Though every common's gone & though traps are set to kill
The little homeless miners.

These are real moles, but 'so close is the mutual ecological
imbrication of the human and the natural that each might stand for
the other'. And this is why, reciprocally, the commons 'worked' for
nature: on them humans lived in the same system of mutually
considerate coexistence as their fellow creatures.

*

But out in the uncommon world the gamekeepers' games
continued. Twenty-four hours before the spring equinox, the West,
always eager to get its first shot in before nature, declared war on
Iraq. Polly and I decided on a big walk into the spring the next
day, all the way from the fens to the Brecks. It was a warm, bright
morning, but the wind was blowing hard from the west, straight in
our faces. It kept the birds down, and, maybe, out – except those
we were primed for, the woodpeckers and bumbarrels. On the fen
the conservation team were still working, cutting back more trees,
dredging out pools, building banks – too much, too late. There
were a few celandines and marsh-marigolds ('Molly-blobs' round
here) tucked into the damp peat by the river, but not much else, so
we took off on unmarked routes, along the ditch edges, under
hedges, looking for a different angle on things, even if it was only
on the wind. There's a botanical measure known as an 'isophene',
which is a line connecting the sites where the average first flowering
of a species occurs on the same day. So the primrose isophene for
21 March might join the Pembrokeshire cliffs and north Devon
lanes and wild gardens in central Norwich (though not, that day,
the Waveney Valley). Isophenes chart the progress of spring across

the country, as it eddies erratically north and east like the edge of an incoming tide. Theoretically it marches forward at a slow walking pace, and given our brisk westing, we ought to have been hitting it full-tilt at 5 m.p.h. But land (and plants) are less tidy and predictable than that, and local topography – a sunny bank or frost hollow – makes its own rules. And in any case, another kind of topography makes these little local adjustments irrelevant, not to say non-existent.

We were blocked by the duck and poultry-farm enclosure just beyond the fen. It was, I guess, half a mile square, a gulag embattled by high-security fences and guard dogs. Drifts of waste plastic blew about the windowless forcing sheds, and wrapped themselves around the wisps of ornamental saplings that had been planted in an effort to hide them. This, it was chilling to remember, was a place where one of the backbones of the national diet was produced. It's striking how the architecture of institutional imprisonment is the same for all nations and all species. And remarkable that it should be tolerated in a democracy in the twenty-first century.

Beyond the battery, the iron grip of industrial farming was deadening. We passed through what, in the nineteenth century, had been the common fens of Hinderclay and Blo' Norton, now reduced to thin shreds of reed-stands and the remnants of pools. The Little Ouse that feeds them has had most of its natural margins and meanders straightened out. Extraction by the water board and local farmers has lowered the water table drastically, and the dried-out peat is being invaded by bracken and birch. Further on there were scenes of dispiriting dereliction: abandoned combines, sprouting hay-bales, piles of old tyres and scrap metal, and a pervasive stench of fungicides and chicken shit; Cold Comfort Farm meets the Common Agricultural Policy.

But the badlands ended. We passed onto sandier soils, on the very edge of Breckland. On the north side of the river, in the tiny parish of Gasthorpe, was a large sheep-walk, a vision of what the whole of the area must have been like at the start of the nineteenth century. There were lapwings swooping overhead, and the first brimstone and peacock butterflies. Hares were playing in the field, leaping and boxing and somersaulting. Three were playing what I can only call Spot the Lady. The front animal in the lolloping chase repeatedly spun round in a full circle, so that it seemed – though I couldn't be sure – that it then became the second, and so on down the line. I remembered a wonderful Middle English poem about the hare, 'the way-beater, the light-foot, the sitter-still, the one who doesn't go straight home'.

And then, impossibly it seemed, we heard the lilt of calling curlews, somewhere out beyond the sheep, away in the pale grassland. In just a few hundred yards, the East Anglian prairie had turned into the Dales, or a salt-marsh, or perhaps just itself, as it was always meant to be. This was no longer a common, but some ancient isophenes, some lines that still joined special moments of natural and human celebration, were moving through it.

4

The Naming of Parts

And this you can see is the bolt. The purpose of this
Is to open the breech, as you can see. We can slide it
Rapidly backwards and forwards: We call this
Easing the spring. And rapidly backwards and forwards
The early bees are assaulting and fumbling the flowers:
They call it easing the Spring.

from 'The Naming of Parts' by Henry Reed, 1941

IT MUST BE SPRING. The gale has switched to the south-west, and it's hard to walk upright. In the brief glints of sun, the guinea-pigs press themselves against the wire-mesh of their cage, braving the wind for a few minutes of warmth. It is all I can do not to set them free. The cats have gone back to lounging in my room, the one island of warmth in the house. When Blanco, always the outdoor boy, comes back through the cat-flap like a meteor, I look out of the window, and see a big dog fox trying to get in, too. His fur is standing on end, as if he had been the one who'd been shocked.

The war blunders on, and rattled by the weather as well as fears of terrorist reprisals, the countryside has gone into siege mode. The supermarkets have put bottled water on ration. Our DIY store has sold out of masking tape, which seems to have become a kind of talisman, believed to have the power to ward off unmentionable plagues. What a hope. An archaeologist friend excavating the remains of a medieval hospital in Scotland (the scene of earlier tribal wars) tells me he has found anthrax spores still viable after six hundred years. And every village has just one house with a Union Jack in the garden.

But the plants are indomitable. The 'blackthorn winter' starts in the second week of April, frosting the hedges while the wild plums are still in blossom. I'm finding flowers and flourishes that are quite new to me. Down in the fen there are the beginnings of things that I can't hope to identify: the first thin spikes of sedges, the russet spears of leaves not yet touched by chlorophyll, finials of the peat. There are precocious cowslips everywhere – on roadsides, churchyards, uncut back lawns. One patch is growing with luminous blue ground-ivy on a dead log – an extraordinary colour combination. Something has happened, some subtle change in climate or soil chemistry, to help the cowslips back. There are newcomers, too. The first big wayside flower of the year here is the white comfrey. It's only been 'loose' here a couple of centuries, naturalised from Turkey, but its starched-linen, wine-goblet flowers are as cheering as snowdrops – which are still in flower, too, three months on from their first appearance.

But I am bothered about the migrants. I haven't heard a willow warbler or a blackcap, normally in full chorus by early April. Not a single passing swallow has hinted at a change in the seasons, and I worry again about the ominously expanding deserts and blighted marshes of the Middle East, through which so many of Norfolk's summer visitors must pass. But in the house we get things ready for them. The barn doors are left open. The windows near where the house martins nest are shut. Kate puts up a painting of a martlet just inches away from the back of their last year's mud-shell. We wait – and on the date the cuckoos should have arrived, the West ousts Saddam.

*

In the mizzling weather I take to walking in the dusk. I suspect there's a trace of asceticism in this, a desire to confront the awful sterility of East Anglia's farming landscape head-on; and maybe, too, a sulky determination to make the worst of this seemingly endless false start, to not see what is not yet there. The dark strips things down, lays them bare in the same way that snow does. All the ornamentations of the daytime, the frills and spurts of spring, are clouded, and you are left with the basic imperatives of geology and weather. And in a ruthlessly agricultural landscape like East Anglia, with the contours of power. In the half-light the edges of the lanes melt away, and the remains of the hedgebanks and ditches that were created two thousand years ago now seem little more than parentheses round the fields. The few patches of woodland, kept for pheasants, are indistinguishable in silhouette from the battery farms devoted to another ex-Asian game bird – except that the low forcing-sheds glow dimly from twenty-four-hour lighting systems designed to ensure the chickens stay awake, fattening-up.

Sometimes I catch glimpses of other dusk-haunters: a few spring moths, a late departing woodcock, pipistrelle and long-eared bats, flickering shapes in that brief window between it being too light for them to fly and too dark for me to see. One evening, on Mellis Common, I saw a jack snipe crouched under a molehill, mouse-like and tense. I thought it was a skylark until it jinked off silently into the twilight, bound for its nesting-ground in the Arctic. We were ships passing in the night.

But so much that was once part of the experience of the April dusk, even in my childhood, seems to be missing. The gnat swarms are smaller, the chorusing blackbirds fewer. Most painfully of all, there are no barn owls. I've heard rumours of one that haunts the western reaches of the valley, but in six months haven't seen a single bird within 20 miles. Up till fifty years ago they'd been parish familiars throughout England, keeping their pale vigil over the pastures. When I was a child a pair had nested in a barn not more than 300 yards from my house, and their hunting range matched up almost exactly with the boundaries of our gang's territory in the Field: over the old brick piles that were all that remained of the Hall, up the ivy-clad wall that lined the council estate, across the steep field we used for tobogganing, then down, if we were lucky, through the bosky edges of our back gardens. The memory of the owls beating past the poplar trees – burnished golden wings against lime-green leaves in the evening light – is one of the few visual images of childhood I can recall with absolute clarity. Now I see the white owl not so much as an object in the landscape but as a creature looking at me. It is an uncompromising gaze: this is *me*, it says, on my estate, about my business. What's yours? No wonder barn owls were seen as guardian spirits, beaters of the bounds between humans and the wild.

Their image has always been an ambivalent one. The tawny owl's generic name is *Strix*, Latin for witch, and there are stories of the Church burning owls for witchcraft in the Middle Ages. Yet in the countryside barn owls (from a related family) were also white witches, symbols of luck and continuity, and farmers would nail owl corpses to barn doors to frighten off evil spirits. In eighteenth-century Selborne they nested under the eaves of the parish church, and Gilbert White noted how they had adjusted their hunting flights to the architecture: as 'feet are necessary in their ascent under the tiles, they constantly perch first on the roof of the chancel, and shift the mouse from their claws to their bill, that the feet may be at liberty to take hold of the plate on the wall as they are rising under the eaves'. One of White's friends in Wiltshire told him about 'a vast hollow pollard-ash that had been the mansion of owls for centuries'. He'd found a congealed mass that turned out to be 'a congerie of the bones of mice (and perhaps of birds and bats) that had been heaping together for ages, being cast up in the pellets out of the crops of many generations of inhabitants'.

Seventy-five years later John Clare, adrift in his asylum, remembered the normalcy, the neighbourliness of the parish owl:

> Now the owl on wheaten wing
> And white hood scowling o'er his eyes,
> Jerking with a sudden spring
> Through the three-cornered barn holes flies.
>
> from 'Evening', written on 14 February 1847

In the early years of the twentieth century they were still hovering in some borderland between the familiar and the mysterious. The transactions of the Norfolk Naturalists Trust carried an extraordinary account of a pair seen in February not far from this valley, and seeming to the observer to be 'luminous'. There was a light mist in the air, and they were floating like will-o'-the-wisps over a patch of marshy ground. One 'emerged from a covert about 200 yards distant, flying backwards and forwards across the field, at times approaching within 50 yards of where I was standing . . . It literally lit up the branches of the trees as it flew past.' The owls had probably picked up phosphorescence from roosting in the crumbling 'touchwood' of trees smitten with honey fungus. Yet they were an eerie enough sight to convince one Norfolk naturalist that they had the power to generate their own illumination. Barn owls are creatures not just of the geographical margins but the cultural margins, too, of the debatable ground.

In East Anglia they call unkempt and unresolved corners of land 'muddles'; and it was the post-war un-muddling of the landscape, the concerted drive towards tidiness and efficiency, that marked the turning-point for the barn owl. Almost everywhere the green lanes and road verges were overmown and drenched with chemicals. The pastures were converted to arable and the stackyards to silos. The barns themselves were flattened, or made into smart houses. Many of the owls were flattened, too, hunting along old routeways that had suddenly become major roads. By the 1990s the UK population was down to less than five thousand pairs, about a third of the numbers half a century before. And about that time there was an uncanny rerun of the excavation Gilbert White had reported in the 1770s. Less than a mile from White's village, an old chimney stack was opened for the first time since it was capped in 1913. Inside were three sackfuls of perfectly preserved and desiccated barn owl pellets. They revealed an immense variety of diet. Among eight hundred identifiable food items were fourteen species of mammal – including water shrew, Natterer's bat, weasel and dormouse – fragments of frogs, swallows, yellowhammers and an abundance of insects. Nothing like this variety has ever been found in modern barn owl pellets, and these relics, preserved in their original site like fossils, were a reminder of the diversity of the countryside before the Great War.

What is one to make of the decline of the screech owl – Norfolk's Billy Wise, Yorkshire's Jenny Howler, Sussex's Moggy – a passing perhaps unmourned by the increasing numbers of people who have never seen one in the wild? Few birds are so dramatically beautiful, or can bring the exquisite delicacy of flight so close to us, or can look at us so penetratingly, eye to eye. But they mean more than that. Ecologists look to the condition of 'top predators' as a measure of how well the ecosystem on which they depend is working. The barn owl is a cultural indicator, too. We recognise, at a deep level, the meaning of that ritual crossing of the fields. It is a sacrament, a consecration of 'good ground' and the boundaries between light and dark, of the proper order of things. Just as the summer migrants stand for renewal, so the barn owl stands for continuity, and its passing leaves us that bit less grounded.

*

But insidiously, teasingly, the migrants began to flit into the valley. It was like the eddying of a piece of gossip – 'There's news! We're back!' – first in one village, then the next, then back again with a new twist.

15 April: Kestrels and sparrowhawks are doing display flights high above the meadow. I surprise two kestrels in the barn, above a little mound of vole-fur pellets. Later that day, there are three swallows on a wire at Dickleburgh, 6 miles away.

16 April: A beautiful, hot day, after a light frost. The wild cherry is in bloom, and I see one swallow at a farm half a mile up the road. The cats are suffused with the rising sap. On my bed, just after dawn, Blackie stretches theatrically out in front of Lily, asking to be washed. Lily obliges, nibbling her face and neck to do the job properly. From the other side of the bed Blanco watches intently. His head is jutting forward, in a tense posture that usually means jealousy, or peevishness. I think he's going to pounce. But he pads slowly towards Blackie and Lily until he's about 6 inches away. Never dropping his stare, he begins flexing his back legs up and down, in the same kneading action that cats use when doing 'contentments' with their front paws. He has an orgasm with a small curdled cry, then pads back to his side of the bed, washes his willy and goes to sleep. Outside, somewhere on the far side of the Ling, the first cuckoo calls.

17 April: Still fine and warm. I slip into the grounds of the big house to see if there are swallows and martins over the lake. There aren't, but four little gulls, *en route* from the Mediterranean to the Baltic, are wafting over the water, the reflection of their black underwings rippling on the surface. Later, I go down to the little fen at Roydon. This is an improbable oasis, 40 acres of swamp and carr within sight and sound of the main road to Diss. Things are showing: the first quill spikes of reeds, the fretwork leaves of meadow-rue, and deep among the alders, sheaves of yellow iris blades. I try to stop prying so anxiously, to relax my ears, let the place wash over me. The fen feels as if it has been hung up to dry in the sun. Last year's sedges look parched, as if it were the end of summer, not the beginning. A tree-creeper appears at the foot of a willow, jerks up the trunk, probing the crevices with its bodkin bill, then swoops down to the foot of the next tree, 'weaving the wood together' in Paul Evans's words. The greenery is burgeoning. There are plants in their aboriginal home here that I only know as hedgerow oddities. Wild hops are shooting out of patches of damp loam, often yards from any support, and beginning to twine up flag blades, tussocks, even each other. Native redcurrants are growing in pools and flashes of stagnant water. Their young leaves have a purple sheen, as if they have been dabbed with wine, and the first tassels of yellow-green flowers, five-petalled, have the look of tiny medieval carvings. The fen is full of more humdrum plants, too — nettle, woody nightshade, the asparagus-like shoots of horsetail. Fenland's intrinsic shiftiness and changeability have made it one of the foundries of the vegetation of disturbed ground, and the species that evolved in its oozes of silt are among those that have found congenial homes in the mud of farmyards and the rich loams of vegetable gardens.

I walk on into the fen. Somewhere above me I hear a snatch of a muted, thrush-like song, and see a bird fly up to the top of an alder. But it's not a song-thrush. It has a dark eye-stripe, and through binoculars I can see it's a redwing. Another joins it, and they sit stock-still, looking south. Then the new bird turns slowly round to gaze at the singer. Its underwing covers show with the same warm russet colours as the alder buds. Is the urge to fly back to Scandinavia stirring in them? Or is the perfection of the light and warmth, the imminent leafing, making them pause? What does it take to make a bird override the hard-wiring that determines migration? Pulled by two contrary instincts, to settle and migrate, would they experience the anxiety of choice? Redwings have bred in England, in Kent, and stayed deep into the summer in Suffolk. In the Chilterns, on almost the same date as today, I once saw a departing redwing in the same hedge as a newly arrived redstart. Redstart, redwing, red-budded tree, redcurrant: so many messages from the shade. But when the redwings flew off the spring fen fell silent, and I thought uncomfortably of Rachel Carson's prophetic book.

18 April: At 9 a.m. there is a flash of white rump and a snatch of bony chatter outside my window. A single house-martin swoops up to one of last year's nests, clings to the shell, then flies away as abruptly as it arrived.

And it was at this moment, out of a blue sky, not the looming clouds of winter, that my old habit returned. I sunk, just for a few days, into a state of free-floating anxiety. It wasn't panic, or anything debilitating, just sufficient agitation to turn my mind in on itself. From worrying about the migrants, I began worrying about worrying, cursing myself for spoiling my enjoyment of what was happening in this first new season in East Anglia. All around me the spring was detonating with ditch-deep primroses and vast congregations of rooks and the delectable hovering furriness of bee-flies, and I was obsessed with these few thin threads of new life that are supposed to wind up from Africa.

Once begun, a spell of anxiety is hard to break. The interruption of attention to the outside world creates a strange sense of unreality, of disjunction. You receive familiar images, but your concentration is elsewhere, focused inside your head. It's often described as having a glass wall between you and the world. Yet it can be a lens too. The wobble it gives to normal, unconsidered perception can lead to eerie heightenings of awareness. Once, out walking in a deep-set anxiety state, I realised that my eyes could resolve the outlines of individual birds a quarter of a mile away. It felt like an unearthly and miraculous gift, and gave me the same feeling of queasy oddness that I used to have as a child when I thought about the improbability of being born at all.

But this time I had an inkling of why this kind of anxiety happened, and knew that that would help it go away. There were all kinds of rational explanations. The significance of the martins, for instance, which had once reliably nested on my old home in the Chilterns. Then more solid concerns. I could quote Ted Hughes' poem ('they've made it again ... which means the globe's still working') and ponder the dreadful implications that lurk between the lines if they ever failed to make it back. I knew that migrant birds, crossing warring nations and increasingly unstable climate zones, had become our modern miners' canaries: they weave the world together too. But I also knew my worries were more personal than this. The migrants set my clock right, reassured me about when and where I was. They were my version of the Christmas tableau.

More mundanely, I know I'm missing my wood, and beech trees, and bluebells. They were my companion plants back in the Chilterns, but don't occur here in the new-growth woods or the alder and willow carr at the edges of the fens. What we have instead, edging into the hedgebanks, is the hybrid between the English and Spanish bluebell, a favourite cottage garden plant which has escaped and become naturalised. Like many hybrids it's more vigorous than either of its parents, and is most definitely 'being about'. There had been a few patches on the edge of my old wood, interbreeding harmoniously with the natives and generating a swarm of hybrids that were beginning to colonise some of the bluebell-free areas. I was rather fond of them, especially an odd sport with stunted, green flowers. Aesthetically they're no real match for the English species. They're paler, stiffer, less subtle – more Laura Ashley than art nouveau. But I like the little tufts and accents and spikes of brightness they lend to the unmediated wash of a lake of bluebells. And, up in this part of East Anglia, they are one of the few splashes of early springtime blue to be had.

But this spring the Jeremiahs have it in for the Spaniards. Believers in steady-state ecosystems and 'the integrity of species' have begun a myth that the aliens will 'hybridise our English bluebells out of existence' – a familiar line of argument to anyone who lives in an inner city. Just what it might mean in the case of plants, and whether such an exotic route to extinction is even possible in the real world, is not at all clear. Our two oak species, for instance, English and sessile, have been cross-breeding freely for ten thousand years without the slightest sign of the one eliminating the other's 'pure stock'. (And how, except in the recesses of a genetics lab, would we ever know?) Nature itself has scant regard for the purity of species, and has been experimenting with new combinations and launching mongrels on the world ever since life began.

But there are good reasons for keeping a watchful eye on organisms that (often with human help) have skipped the slow testing of evolutionary time, jumped habitats and naturalised themselves where they were never meant to be. Warmer parts of the world are full of cosmopolitan species that have usurped the less adaptable natives. But it isn't sensible to take rigid, ideological stances towards nature, which is intrinsically pragmatic and quick-footed, and for every 'alien bully-boy' (as one conservationist called them) there is a benign opportunist that has filled a blank space, colonised a new island, taken advantage of an interregnum. We should be grateful for them. In the confusion of man-made climate change, as vulnerable native species like the bluebell begin to feel the stress of rising temperatures and unstable seasons, new, more adaptable species will begin to fill those vacuums that nature abhors.

Maybe even this balmy April is man-made. On one particularly transcendental morning, I watched a wildlife trust spokesman discussing the 'threat' posed by the Spanish interlopers on the breakfast news. It was the kind of day that makes one feel like saying grace for a new blade of grass. But he stood in his Home Counties reserve and declared that he'd 'stamp on' any foreigners he saw.

A mite of resentment was added to my migration anxieties, and wishing misery on all managers, I stumped off on a plant twitch to Wayland Wood, where the legendary yellow star-of-Bethlehem, *Gagea lutea* grows. If the miracles of spring refused to come to me, I huffed, then I was going in search of them. It was about as sensible as running away to sea. I'd failed to track the plant down in this spot twenty-five years before; and in any case, didn't so much want to see it, as find it, make a small verification of the continuity of things and, more to the point, I suspect, prove that my incomparable skills at plant dowsing were still intact.

Wayland, a half-hour's drive north of the valley, is one of Norfolk's few surviving ancient woods. Its name comes from the Old Norse *Wanelund*, meaning a sacred grove. This was a place of assembly, maybe of pagan worship, long before the Conquest. Later, medieval child abuse in a local manor house made it the site of the Babes in the Wood legend, which produced an alternative derivation of its name. It was the Wailing Wood, where the cries of the abandoned children could still be heard at night.

The yellow star-of-Bethlehem is itself a thing of ancient and rather sad echoes. When I first came looking for it here I'd been lured partly by the poignant account of the only occasion on which it had been seen in my home county of Hertfordshire. This featured another babe in the wood, a girl from Ware Grammar School, who, our county flora ruefully reported, 'found it in Broxbourne Woods, but could not recall the exact location'.

A quarter of a century on, I was no more clued-up myself. I'd still never seen the plant, and knew it only from illustrations in books, where it looked rather like a diminutive and unprepossessing celandine. I had no idea of its jizz, or the sort of spring weather that might overcome its notorious disinclination to come into flower, or even the kind of woodland niche it favoured. I was a trufflehound without its nose. So I quartered the wood, tacking across its mysterious earthworks. I peered round all its inner recesses, in the coppice glades, along the edges of the rides, in the patches of dapple under the taller trees. I tried to be clever and nudge my way down tracks that had brushwood piled in front of them, thinking that this might be a deliberate deterrent to *Gagea* stalkers. But I knew I was hunting blind. I didn't find a thing. I was a stranger and a train-spotter and it served me right.

When I gave in, and began just meandering about, the wood became enchanting, spangled with wide-open celandines and the first early purple orchids. There were wood anemones, too, in an ethereal shade of rose. Where they were growing directly under the trees they created little cameos of almost Japanese simplicity: the pale petals, as sharp-cut as paper flowers, the moss on the bare earth, the dark trunks behind. And coming into flower above them was something that, to tell the truth, was more special to me than the *Gagea*: the white, flour-and-marzipan-scented racemes of the bird-cherry. I was, in a trice, back in the Yorkshire Dales in June 1985, saying an emotional goodbye to a close-knit film-crew after a long shoot in desperate weather. The bird-cherry was still in flower after the icy spring, but quite leafless from ermine moth caterpillars; and the great sugar-loaf of Ingleborough Hill loomed in the distance under a storm cloud we'd escaped by minutes. Every time I see bird-cherry now, I remember that day, and the moths' cocoons, and how we beat the bad weather in the end.

When I'd been in Wayland before, I'd met, by chance, the last freelance woodman to work there. He'd been cutting brush near the main track, over the few acres he'd bought coppicing rights for. He made hurdles from the hazel and broom handles from the ash. The bird-cherry wands, he assured me, made perfect chrysanthemum stakes. He was nervous about his future now the wood was being converted into a nature reserve, but had noticed a revival in the wood's fortunes. He pointed to the early purple orchids sprouting in the track. He called them 'cuckoos'. He'd never seen the star-of-Bethlehem himself.

*

Our summer idylls in the Cevennes were full of name-games and recaptured experiences. After so many trips there we had our own special rites of moment and place. Each morning, on the way back from washing, we'd listen to the scrabbling jingle of the local serin, nicknamed *la chanson de la toilette*. We'd make wasps' breakfasts out of saucers of jam, to keep them away from the croissants. In the afternoon we went swimming in the River Dourbie and, breast-stroking half underwater, necks as stiff as swans, seemed of no more concern to the river's other citizens than the lumps of polystyrene that occasionally floated down. Crag martins dipped for water in front of us. Young viperine snakes wound past, snatching delta-winged flies from the surface. Banded grayling butterflies, as big as willow warblers, flew in procession overhead. We chatted about butterflies as we lolled in the heat, and, as the array of species began to resemble the cast of a Greek drama (there were Cleopatras, Satyrs, Dryads), so, doped silly by the heat, we began to fantasise our own list of theatrical sunflies: the True Blue, the Bent Copper, the Grey-rinsed Skipper (a.k.a. Old Heath) and the Large White Supremacist, *Eugenea terreblanchea*. This was alternative humour showing its middle age – but then all those classical tags were invented by middle-aged pedants on their summer holidays, too. In the evenings I dozed outside, listening to the poignant creaks of my friends' children in their hammocks. They were edging into adolescence, and already beginning to put their own twists of irony between themselves and this place. One morning they made an *un*-nature trail, hanging strings of garlic in trees, and making bird-of-paradise droppings out of striped toothpaste.

But the most durable tradition was the evening plant salon. We'd sit with our field-guides and our ragged bundles of leaves and flowers spread out on the table, and talk them through, taking every opportunity to be side-tracked. Orchids (though we never picked these) had a special fascination. Many had the look of elaborate conceits in porcelain, or colonies of hatching insects. The botanists who named them saw likenesses too; to lizards, bees, bugs, butterflies, spiders, even pyramids. But in the tribe (*Orchis*) with which we were most familiar, they saw chiefly little homuncules, formed from the heads, or helmets, and arms and trailing legs of the individual flowers. The size of the hoods, the sinuousness of the limbs, the waisting, the elegant drape of the lobes, made them variously men, soldiers, ladies and, in one particularly gangly armed type, monkeys. But we could scarcely tell them apart. The hills were alive with unresolved manikins and hermaphrodites. Orchids are one of the most recently evolved groups of plants, still hazy about their identity, and they hybridise quite promiscuously.

Sometimes we'd dine out at Le Papillon in St Jean du Bruel, a botanical experience in its own right. We'd eat wild asparagus, and last year's wild mushrooms from the sheep pastures, and local honey flavoured with wild thyme. One spring a German photographer was staying at the hotel attached to the restaurant, and had left two thick volumes of orchid photographs on the bar for customers to browse through. He'd managed to track down most of the west European species in the country between St Jean and the *causses*, and his pictures were a tribute to his attentiveness and understanding of the landscape. But it was his exploration of the diversity inside each species that was most intriguing. His albums were a toast to idiosyncratic local forms, to upside-down pyramids and wingless bees and white ladies. These were the *orchidées du pays*, as tangy and particular as the local wines. It was a relief to see that someone else shared our enthralment to these shape-shifters. Yet some atavistic urge to name them, to pin them down, still nagged at us. One variety in particular, the rare and almost mythical cross between the man and monkey orchid, was weirdly gripping. We nicknamed it the missing link orchid, but doubt that we were the first.

I wondered about these word-games sometimes, but they were our specific – and tribal – way of relating to the Cevennes' other inhabitants. In those weeks in the open air, in the company of dozing beavers and cavorting choughs, we loosened up, lost our self-consciousness – or perhaps, more accurately, just accepted it. We were stalking and sniffing and savouring with the rest of the locals, but don't think we had any pretence of going feral. Our intellectual fantasies and lamentable jokes were our way of playing, of joining in a relish of the sun that is, paradoxically, older and deeper than language.

*

Now, even when I'm botanising by myself, I find I still want to put a name to things. It seems to be a basic human reaction, the first step in beginning a relationship: 'What's your name?' But down in the fens it's proving a struggle. I squint at plants far out in impassable reed-swamps, at plants not yet in flower, at sedges which are, but which might as well be Rubik's cubes for all I can deduce from their tangled inner parts. Most of my books are still in that north London storage container. I'm rusty after two seasons out of the field, and frustrated that I don't know what's what.

So why bother, an unfamiliar, relaxed voice inside me says. Why not simply relish the spring's new life (and *your* new life, for goodness' sake) – its exquisite variegation, the interplay of the yellow moss-ground, the filigree sedges, the solid mass of tussocks, the growingness of it all? Well I can, I think, but don't find it easy to stop there. Some inner tic – not just an intellectual reflex – makes me want to know who they all are. Maybe it's just a hangover from whatever impulse makes boys collect stamps. But some kind of naming is a prerequisite of talking about plants, and I'd make a strong argument in favour of its cultural (let alone scientific) importance. Some years ago, when he was toying with Zen Buddhism, John Fowles suggested that 'the name of a plant is a pane of dirty glass between you and it'. I've never been able to share this feeling, even though I understand what he was getting at. It seems to me that naming a plant, and for that matter any living thing, is a gesture of respect towards its individuality, its distinction from the generalised green blur. It is, in a way, *exactly* a gesture: as natural and clarifying as pointing. The kind of name – scientific, popular, fantastical, pet – scarcely matters, provided someone can communicate it.

The historian Maria Benjamin once described natural history as 'ideologically loaded housekeeping', and its preoccupation with naming and ordering as 'taxonomising the world's bounty into a pattern of strict hierarchy'. And 'the naming of the beasts'

(Adam's first piece of housekeeping) was of course the crucial groundbase for the modern world's project of appropriating and taming nature, of turning it into an object. But that was a consequence of what you might call the ecology of naming, of the culture and view of nature it emerges from. In itself naming is no more colonial or 'capturing' than cave painting.

I've been browsing for local names for those Wayland Wood plants, yellow star-of-Bethlehem and bird-cherry, wondering if they might reflect how these plants were seen and thought of in East Anglia. In Lincolnshire bird-cherry was the mazzard (more usually the wild cherry); in Yorkshire, hagberry or hackberry, from the old Norse, *hegge* or *hagge*, meaning to cut or hew. Maybe this was oblique reference to the bitterness, the 'edge', of the cherries, or to the fact that the shrub was cut as coppice, or then again flourished when larger trees were cut. Maybe, given the reluctance of language to stick to the point, all three. But here no local names have ever been written down. Perhaps, practically, bird-cherry was just one of a general group of 'brush' or 'spring', or stick-plants. Groupings and taxonomies of this kind are common in tribal and peasant cultures, and still persist with us. Species have been grouped by use, by edibility, by seeming gender affinities, by season, by 'heat' or bitterness, by demeanour. Most commonly, they are grouped by size. Scientific botanists insist, quite correctly, that both grasses and trees are varieties of the general order of flowers. But in vernacular opinion across the world, they are clearly seen as different sorts of plants, separated off not just by their stature, but by their occupation of different layers in the landscape.

Even scientists use functional and holistic taxonomies at times. Perfumiers group together biologically unrelated species which share the same aromatic compounds. Ecologists make regular use of so-called 'indicator species', suites of plants that characterise particular ecosystems, and record the way in which soil, climate, time and place are reflected in vegetation. Most of us have our own less stringent versions of indicator taxomonies, groupings of plants that express special moments of the year or favourite spots. The conventional grouping and naming of species according to their anatomy and kinship history is useful, and gives a unique name to each species capable – at least theoretically – of being understood anywhere. But it's no more true or 'natural' than these vernacular clumpings. The hagberry's real-world connections, for instance, aren't with its close relative, the wild cherry, the gean, but with limestone and ermine moths and pied-wagtails – an upland bank-and-wall taxonomic group. In other words, an ecosystem.

John Clare, a respecter of the privacy and identity of wild things, was meticulous in the language he used to identify them. His publisher once queried Clare's use of a dialect name for the insect known as the froghopper. 'Woodseers', Clare tetchily replied, 'are insects which I dare say you know very well

> whether it be the proper name I don't know tis what we call them & that you know is sufficient for us – they lye in little white notts of spittle on the backs of leaves & flowers. How they come I don't know but they are always seen plentiful in moist weather – & are one of the shepherds weather glasses. When the head of the insect is seen upward it is said to token fine weather when downward on the contrary wet may be expected.'

'Woodseers' means 'wood-prophets', and Clare was placing the name not just in the context of the insects' own 'culture', but of the larger community of shepherds and other weather-sensitive creatures. Just as scientific naming aims at differentiating species over a baseline of kinship connections, so vernacular naming attempts to show similarities and associations over a more broadly based system of differentiation.

But the price (if it be such) of emphasising connectedness is ambivalence, as Clare found at the end of the woodseer's trail. Froghoppers' 'white notts of spittle' were (and are) widely known as cuckoo-spit. Cuckoo-spit was also the northern name for lady's-smock (*Cardamine pratensis*, the 'meadow-cress'), which much of the rest of England, including East Anglia, referred to as cuckoo-flower. Geoffrey Grigson found cuckoo names, either alone or in some combination, for some twenty-five British species, including cuckoo-pint (cuckoo's pintle, or penis), cuckoohood (for corn-flower in Scotland) and cuckold buttons (burdock in the West Country). The names chiefly refer to the fact that the flowers appear at the same time as the cuckoo, but they're also a nest of *doubles entendres* and puns about the goings-on in the meadows where cuckoos sang and ladies' smocks were lifted.

But Clare's loyalties were the same as the Wayland woodman's, and he insisted that the true cuckoo flowers were unquestionably orchids:

> these is my cuckoos & the one that is found in Spring with the blue bells is the 'pouch lipd cuckoo bud' I have so often mentioned its flowers are purple & freckld with paler spots inside & its leaves are spotted with jet like the arum they come & go with the cuckoo & in my opinion are the only cuckoo flowers of England let the commentators of Shakspear say what they will nay shakspear himself has no authority for me in this particular the vulgar werever I have been know them by this name only & the vulgar are always the best glossary of such things.

Clare's cuckoos comprise a group of half a dozen orchids, which he differentiates partly by prefixes, partly by descriptions. But what also particularises them are their locations. The orchids on the common are not the same as the orchids in the lane. William Hazlitt, in his essay 'On the Love of the Country', argued that 'the interest we feel in human nature is exclusive, and confined to the individual, the interest we feel in external nature is common, and transferable from one object to all others of the same class'. This is partly true: affection for primroses, cuckoos, whatever, is felt generally. But it is also felt, by all creatures, specifically for localised plants. Plants are part of what makes a locality, differentiates it, makes an amorphous site into a place, a territory, an address. Clare once called wild flowers 'green memorials', a view echoed (perhaps a shade too optimistically) in Ronald Blythe's phrase 'a form of permanent geography'. Clare's orchids were individuals, located, known, and, like all individuals, vulnerable in a way that their 'class' might not be. In his list of twelve 'English Orchis', he describes three 'cuckoos', and his *Orchis latifolia* (marsh-orchid) 'grows in a low part of Mr. Clarks Close at Royce Wood — was very plentiful before Enclosure on a spot called Parkers Moor near Peasfield-hedge & on Deadmoor near Sneef green & Rotten moor by Moorclose but these places are now all under plough'.

*

It's late April. The garden is full of primroses and courting birds. A family of rabbits, out of a touching if foolhardy sense of security in this arcadia, have dug a burrow right underneath the pear tree. Blackie has followed their every move, and now sits at the living-room window eyeing the youngsters. She has no philosophical problems about being a part of nature and apart from nature. The pane of glass, framing her destined booty, causes her neither confusion nor alienation. She doesn't attempt to burst through it, or stalk defeatedly away. She understands its mediating role perfectly. She simply sights up the young rabbits, moves a pace towards the door, slips back to check their position, then hurtles out the cat-flap, round three sides of the house and polishes off another one.

High above this cultured but remorseless foraging, three great spotted woodpeckers (including two males, I think, from the redness of their napes) shimmy about the pear tree in complicated manoeuvres. The books say that the males fight, and their red tail coverts are often mistaken for bloodstains. But here the chasing and courtship seem to be courteously formal, like a three-dimensional square-dance: up one step, two along, take your partner, keep your spacings.

And at the edges of the lawn, the cock pheasants are strutting and preening too, beating their gorgeous wings so extravagantly that they almost fall over. The females seem to spend all their time in harems and hen parties, and I wonder when they will get round to nesting. But then most of them will soon be dead. Of all the excesses and sacrifices of the spring, the pheasants' fate is the only one that is truly tragic. They are birds deliberately introduced from a far-away habitat, raised in captivity, released prematurely into the wild to be shot, and finishing up, most commonly, under the wheels of cars. The roads around the valley have become pheasant charnel houses, carpeted with their squashed and dismembered remains. I've found severed heads glowering from the tarmac like voodoo warnings; and, in the lane outside the house, a torn-off wing caught in the cow parsley and flapping eerily in the breeze. The air-blown down on the feathers made them still feel warm. This, I think, is what stops me picking up road-kills and taking them home to cook: not that they are so dead, but that, snuffed out so gratuitously, they seem still so very much alive.

Pheasants were introduced to Britain several times from

Asia. The Romans brought over a race (var. *colchicus*), distinguished by a black neck. But, true to their authoritarian form, they kept them in pens, and these pheasants seem not to have gone feral. A thousand years later they were re-introduced by the Normans, this time as the familiar race (var. *torquatus*) with the white neckring. They were adaptable birds of light woodland and scrub in their homeland and naturalised quite successfully in our woods and parkland. They were hunted sporadically with nets, or surprised and coshed when they were roosting, but were basically treated as wild birds. But since organised shooting began two centuries ago they've become Romanised commodities again. They are hatched in incubators, reared in pens, over-fed so that they can barely fly, and then released without ever having learned about danger and territory in the way that young wild birds do. ('Slow down for game birds', reads one shooter's car sticker.) The numbers released each summer are colossal: twenty million in England, and an average of five thousand per estate in East Anglia. They are, during the late summer and autumn, far and away the commonest bird in the region.

A little over half the birds are recorded as shot, but even this is greater than there is any demand for as food, even in rural areas, where they are pretty much disdained. The excess carcasses are burnt, buried or just dumped in heaps in the woods, often only yards from the pens from which they were released. (Pheasant woods bear the scars of feeding and interment in the form of spreading patches of rank weeds among the ground flora, the marks of human extravagance wherever they are found.) The remainder wander aimlessly about the lanes, trying to make sense of the habitat into which, willy-nilly, they've been dumped. As birds whose genetic ancestry predisposes them to running from danger rather than flying, it's no wonder they come off as badly in their confrontation with the car as they do with the beaters and their dogs.

The tragedy of the pheasant lies in the inevitability of its fate, as a displaced and disorientated creature. Collisions between wild animals and the forces of civilisation are, sadly, inevitable. But deliberately to submit a semi-domesticated bird to the same perils smacks of a wilful absence of care, and a particularly callous version of the steward's role.

*

But the migrant birds still seemed elusive and sparse. The swallows of a few weeks before had deserted the farms. I hadn't heard the shrill flutings of the blackcaps that should have been abundant in the fens, or for that matter that first herald of spring, a chiffchaff. Had they been disorientated, too, blown off their traditional route-ways by Mediterranean storms? My nightmare, that those ancient ecological links with the south might finally be broken, wouldn't go away. And, as usual at such times, I whirled about, looking for reassurance. I went over to the Suffolk coast, and heard a single nightingale. I pestered friends, who all seemed to be seeing and hearing more than me. I phoned up Bird-lines, Migrant Watch-lines, and heard that, yes, there was a weather block over the continent slowing movement down, but the birds were coming through. And I began to wonder if it was me, as much as them, that was 'blocked'. My hearing had been deteriorating for the past decade, and I had to face up to the possibility that I had lost most of the high-pitched warbler songs, just as I had lost the screaming of swifts. It wasn't a pleasant prospect, being cut off from the one thing that made me feel *not* cut off. Gilbert White had bouts of deafness in middle age, so that he lost 'all the pleasing notices and little intimations arising from rural sounds'. 'And Wisdom', he quoted at the end of a letter, 'at one entrance quite shut out.'

That late April Polly and I went over to the Broads again, to see our friends Mary and Mark Cocker. Mark is an enviably acute man, as sharp in the field as he is in his writing, and I knew that a walk with him would settle things. We make up a bit of a mixed flock, three adults, Mark's eight-year-old daughter Miriam, and a neighbour's son, Kevin, who has insisted on tagging along. We walk along the River Chat, by reed-fringed ditches and clumps of willow scrub. The water flashes intermittently a few paces away. Mark spots sand martins while I'm peering at the ground. He picks up the whiplash song of a Cettis warbler, something I know well and was sure I was capable of hearing. He seems able to conjure birds out of thin air. Then, smartingly, he spots swifts high up over the mere we're walking towards. This was not how I wanted to see my first birds of the year – my first proper swifts since I'd been ill – pointed out for me while I'm looking the other way. I feel humiliated, and hurt, as if I had had a personal gift unwrapped for me. The birds were here, and I was simply failing to register them. Was my hearing even dodgier than I thought? Was my attention still on that wall that I gazed at for so long?

I watch Mark. He's standing upright, scanning forwards, but still holding hands with Miri and talking to her. He calls out terns, a marsh harrier, more martins, which I'm simply not picking up. So I watch myself, too. I'm conscious of the fact that my mind is constantly drifting. My eyes are pointed downwards, focused on some important spot of ground about 5 feet in front of me, a crass compromise between staring nervously at one's feet and navigating forward. It's a botanist's stance, but a depressive's hunch, too. Birders use the jargon word 'dipping' for failing to see rarities, and it seemed a peculiarly apt metaphor for me at that moment.

Kevin, meanwhile, arms akimbo, is talking gibberish about

computer games. He hurls stones at the water, drops the binoculars I've lent him, tears up handfuls of reeds, nearly falls in. He's got attention-deficit syndrome. Miri shakes her head at him, and tells me about her elder sister Rachel's forthcoming appearance in the school production of *Wind in the Willows*. She's playing a weasel, and Miri makes the battle for Toad Hall sound like a piece of woodland agitprop, as the oppressed creatures storm the seat of privilege. I think I'm suffering from attention deficit too. I'm projecting my own debility onto the natural world, in a bizarre physiological version of the pathetic fallacy (that tendency among the Romantics to see human emotions reflected in the workings of nature). The migrants may be late, and down in numbers, but it's me that's really missing the boat, and the cues.

The frustration and sense of loss – how many more springs had I got left? – drove me into a rare bout of technological busyness. I could do something about my attention, but for my hearing I needed help, some clever escape route out of my personal silent spring. I went to see my audiologist, who denied there was anything superior to my hearing aid. I thought of the ear-trumpet that was among Gilbert White's effects, but felt such a thing might be a tad ostentatious for the fens. David Cobham did some lateral thinking and suggested I went to a detective agency, but they informed me that portable devices for tuning in to distant sounds were just so much film-makers' fiction. So it was down to me, and a handful of specialist electrical shops, and the end-result was a combination of a high-quality directional microphone, a digital voice recorder and a pair of Walkman headphones. I called it Auric. The first time I took it out was a revelation: the birds I thought might be stretched out between Baghdad and the Alps – gassed, starved, gale-wrecked – seemed to be singing exultantly only a few feet away. I heard properly, for the first time since my thirties, the little grace notes in reed warblers' songs, the scratching of whitethroats, and that thin, triumphant see-saw of the chiffchaff.

And, paying for that artificial recapturing of youthful senses, I also heard, enormously amplified, the shattering roar of distant aircraft and the hum of traffic. It was, I felt, a fair swap, since these are the realities of the world to which our migrants return. But they'd made it back to where they belonged, and back, too, inside my head.

*

A few days later we went back to one of the wildest, wettest stretches of Broadland, me with Auric stuffed in the bag I'd worn as a Christmas shepherd, and Polly, I fervently hoped, with her lifelong meanderings through the Broads engraved in her head and feet. We left behind all the paraphernalia of the countryside interpretation business, the hides with gates, the trails waymarked in five colours, the boards that tell you what to look for and what to feel about it, edifices that seem to have converged with the precautionary fences and smoothed-out paths of the new safety culture to convey an ominous common message: 'You are not encouraged to have First-hand Experiences. They may hurt. Life is Dangerous. Keep Out.'

We were in an area where close on 6 square miles of open water and marsh are cut by just a single turn-again road, so we followed old trackways and dyke tops and river edges. There were marsh harriers almost everywhere we looked, glancing over the reed-beds, lifting in thermals, the male birds' wing-patterns of grey and chestnut and black flashing like semaphore signals. At times we could see half a dozen in the air at once. In 1971 there was only one pair nesting in the whole of Britain. Now there were upwards of 150 birds (the males are polygamous) in Norfolk alone. With their cowled eyes piercing up and down and across — how I wished I could do that — they had the look of prospectors, on the frontier of things.

As the sun set, three snipe rocketed out of the marsh. They sped up together to about 100 feet and then burst apart like a star-shell. I followed one of the birds through my binoculars. His wings were fanning, beating furiously. He skimmed the rim of the sun, gaining height, then cut abruptly down, like the swoop of a sword, his tail-feathers spread wide apart. He fell for no more than two or three seconds, but through Auric I could hear, for the first time in twenty years, the bleating made by the wind rushing through his taut outer tail-feathers. It sounded like the thrumming of an arrow hitting the target. Snipe do make vocal calls, little wheezy protests when they are flushed, but this drumming is their spring song, their reed-wind rhapsody, blown together from air and feathers.

The setting sun was enormous. It had turned the marsh tawny, as it had been in the autumn. All the shadows were behind us, and it was another kind of darkening at the periphery of my vision, and maybe (though I am not sure of this now) some susurration at the edge of hearing that made me glance up. Three cranes were flying low almost directly above us. Their 8-foot wing-span briefly blotted out the sun. Their bowed silhouettes — the trailing legs and necks held low beneath the body — were like sea-creatures, or great boats. They were rowing into the sunset. And with huge nonchalance they did not swerve a degree as they passed over us. They had been on this marsh for a quarter of a century and had earned their common rights. They came down a couple of hundred yards away from us, two vanishing behind a reed-stand. But the third, just for a moment, danced. It picked up one foot after another and dipped its head. Then it went, I think, fast to sleep.

The cranes' full dance eludes me, a marshland ritual that perhaps I haven't earned the right to witness yet. Yet it's been a compelling, infectious motif throughout history. A crane dance was performed by Theseus on his return from Crete with the youths and maidens he had freed. It was 'a round dance in crane step'. There was a 'dance of white cranes' in China in 500 BC with a ritual pattern so similar it is hard to avoid the suspicion that the two were related. John Clare describes a game called 'acting the crane' which was played at the Harvest Home feast, and which suggests a strong folk-memory of the time the birds were familiar across eastern England: 'A man holds in his hand a long stick, with another tied at the top in the form of an L reversed, which represents the long neck and beak of the crane. This, with himself, is entirely covered with a large sheet. He mostly makes excellent sport, as he puts the whole company to rout, picking out the young girls, and pecking at the bald heads of the old men.' (It's well known that, with cranes, mimicry is a mutual business, and that the birds' dance can be sparked off by the sight of humans cavorting like this.)

The distinguished ornithologist and folklorist Edward Armstrong had no doubts about the common ancestry of human and animal ritual:

> Physiological and emotional needs, sociability, the urge towards the satisfaction of the sense of rhythm, and the inherent tendency of life to express itself in patterns of behaviour have given rise to similar activities amongst birds, beast and man; thus it is no mere capitulation to anthropomorphism when those describing bird dances use terms such as 'quadrille', 'minuet', 'waltz', 'pirouette', and 'setting to partners', but a recognition of the resemblances between avian and human dance ... there is one generalisation we can make. In the dance, the individual reaches out beyond his isolation and seeks to realise that

harmony between himself and the external world without which neither health nor happiness can be achieved.

Armstrong tabulated the basic, universal forms of dance (Static, Ring, Line, Place-changing, Solo, Dual, Dual-communal, Communal, etc.) and found that the elaborate festivities of cranes fitted into almost every category. But his holistic view of ritual is unfashionable now, and I haven't yet found a full, detailed description of the dances of cranes to make up for what I've missed. But in nineteenth-century America, when whooping cranes were still common, the Florida backwoods novelist Marjorie Kinnan Rawlings wrote an evocative account of their evening ceremonials:

> The cranes were dancing a cotillion as surely as it was danced at Volusia. Two stood apart, erect and white, making a strange music that was part cry and part singing. The rhythm was irregular, like the dance. The other birds were in a circle. In the heart of the circle, several moved counter-clockwise. The musicians made their music. The dancers raised their wings and lifted their feet, first one and then the other. They sunk their heads deep in their snowy breasts, lifted them and sunk again. They moved soundlessly, part awkwardness, part grace. The dance was solemn. Wings fluttered, rising and falling like outstretched arms. The outer circle shuffled around and around. The group in the centre attained a slow frenzy . . . Suddenly all motion ceased. Then the two musicians joined the circle. Two other birds took their places. There was a pause. The dance was resumed. The birds were reflected in the clear marsh water. Sixteen white shadows reflected the motions.

To see the Broadland cranes dance — that most sociable and

hedonistic of animal pleasures – will continue to be a dream. But I doubt if I will ever witness it by conscious effort. It will be when I least expect it, somewhere off my beaten track. A wild surprise.

*

The swifts may have been back in the Broads, but they were not yet in the parish, and certainly not back in the loft above my room. On May Day, when I used to hold my blazer collar for them, I went back to the park lake again. Nothing to be seen – but then, in an old discipline remembered, I tilted my binoculars up, and there they were, a boiling maelstrom, just out of naked-eye view, trawling insect plankton just below the cloud base. Some old romantic yell sang in me, and probably out loud too: 'They're Back! The Swifts are Back!' They'd broken through the weather system – if in fact they'd been held up at all rather than just resting out in their own space. In another great modern swift poem, Anne Stevenson sets the birds free from our expectations and symbolic associations:

> ...... in reality
> Not parables, but bolts in the world's need, swift
> Swifts, not in punishment, not in ecstasy, simply
> Sleepers over oceans in the mill of the world's breathing.
> The grace to say, 'They live in another firmament',
> A way to say, 'The miracle cannot occur'
> And watch the miracle,
> This is the truth of swifts, the gift of swifts.

And then a calm descended. The wind turned west and warmed. A turtle dove purred in the fen. Watching from the kitchen window one morning I saw a cuckoo fly low over the back meadow – quick shallow wing-tremors, trailing tail. No wonder they were thought to change into hawks in the winter. And at half past nine the martins returned. Two pairs this time. They began building on the shells of last year's nests. I remembered this moment from half the years of my life, this promise of hectic house-guests for the summer, and with them comedy, company, enlightenment and, too often these days, disappointment.

I settle down outside to watch them, just as I used to when they nested on my old home. I'm careful not to get too close. They are, in these first few hours and days, nervous and temperamental, apt to desert before they've even started. I squat on the lawn and they flash past me, sleek and buoyant, small aerial dolphins, sometimes misjudging the arc they must follow to lift up under the eaves, and having to make their approach all over again – back over the barns, a dip below the telegraph wires, a double wing-flick for adjustment for landing. Their trips away from the nests are only lasting a minute or so, and I'm intrigued about where they're gathering mud. I creep round the barns on the other side of the hedge, following the general direction of their flights, and find them paddling and delving in the puddles on the sugar-beet stand, only fifty yards from the house. The water's as sticky as treacle, and I wonder what the nests will be like after a few days, baked in the sun like flapjacks.

With half-complete nests to build on, they're team-working already. One bird delivers mud to the rim, the other works from inside, plastering and dibbling it in with rapid hammerings. Occasionally the technique changes, and they use slower movements with their bills open, squeezing the mud pellets into place. They have a difficult curve to cope with, where part of the nest has fallen away from the roof, and they're making and mending upwards from the base of the nest, rather than outwards from the wall, as they do when starting from scratch. They stop work at eleven, presumably to give the new courses of mud time to dry and harden. How do they manage this improvisation, with this odd building material, and an engineering situation they can have had no direct experience of? There is no way instinct alone could provide for all the unpredictable and shifting challenges posed by a task like nest-building. These birds are thinking, working out solutions as they go along. Instinct may provide a kind of vocabulary and grammar of action, but for the minute-to-minute details these birds (maybe only nine months old) have to be, like many other creatures, innovative and creative.

I still have the journal my sister and I made during a martin summer in the Chilterns in the 1970s, a meticulous log of the birds' movements that I'd hoped might build up a picture of the texture of their lives and decision-making. Much of it is diagrammatic, sheets of pictograms showing in what direction the birds entered and left the nest, how long they stayed, how long they were away. I thought some rhythm might be visible among the swirling masses of arrows, but there isn't. Just a sense instead of decisive, intelligent hunches. Sometimes, in bad weather, they were away for agonisingly long periods, and I fretted about the nestlings, and about the looming onset of hypothermia and starvation. One afternoon I cycled off to see if I could find out where the parents foraged. I had a vague image of some kind of communal feeding station, thronging with birds from all over the parish. But I found nothing of the kind, even after hours of searching. Instead there were little packs of birds feeding wherever they could find shelter from the wind. I discovered one bird strafing backwards and forwards deep inside a canal lock, and another group skimming the tops and skirts of a lime tree, brushing against the leaves with their wings to dislodge insects.

There were written notes in the log, too, about a last brood raised in October, when there was a rime of frost on the nest; about raids by a hobby that had my loyalties entirely split, and had me shouting in the road at both species to keep their distance; and about another summer assault by a great spotted woodpecker that chopped away at the nest-hole and threw all the nestlings on the ground. (The parents spent most of that night chattering in deep debate in the wrecked nest, but the next day worked furiously on repairs, building up the hole so tightly they could barely get in themselves.)

This was what I wrote when the chicks flew the nest, the day of the flitting:

> The chicks have been active and excited for the past couple of days, bobbing up and down in the nest, stretching their wings, staring out. The ones at the entrance keep disappearing backwards, and the others wriggle forward for their turn. Sometimes I'm sure they stand on each other. They look like young otters, perched upright at the hole. They must be close to flying.
>
> Today the largest chick begins to hang out of the nest, right up to its chest. It already has almost full adult plumage. Its bristly crest has been replaced by a smooth, sooty cap and a silky white front. It peers round as usual — up at its own shadow, down at the spiders spinning webs around the base of the nest, out at passers-by and cars, and especially at the cabbage white butterflies dithering past the nest. It follows their progress until its head is screwed round at almost a hundred and eighty degrees. Then, suddenly a chick again, its beak gapes involuntarily. It's dreaming of food.
>
> Now the parents are teasing the chicks, flying up as if to feed them, then darting away. They cling to the wall very close to the nest and weave their heads at the

youngsters like snake charmers. One seems almost on the point of popping out when they all cower back in the nest at the sight of a bright yellow British Telecom van.

About lunchtime the biggest chick comes to the nest hole again, stretches its neck forward with a curious exultant churring note, and pops out like a champagne cork. The next is out a few minutes later and they are soon both flying as airily as their parents, dashing back and forth to the nest entrance and chivvying the remaining youngsters.

*

That stage would be some weeks off with our martins at the farmhouse. But before the young had even hatched, I got myself involved in an odd little drama, whose props included martins and swifts and the piece of Gilbert White's yew tree which had found its way up to Norfolk with me, and whose denouement involved some revelations about the 'narratives of nature'. I need to explain something of the background to these improbable connections.

Gilbert White's immortal book, *The Natural History of Selborne* (1789), includes four beautiful and insightful essays on the hirondelles. (I use this obscure word, because, unlike 'hirundines', it includes swifts along with swallows and martins, which in popular taxonomies belong together as aerial summer house-guests.) They are generally thought to be the prototypes of 'objective' nature-writing. It was partly these that led to my own fascination with White, and eventually to the writing of his biography. At that time I often stayed in Selborne, where one of the features of the village landscape was the ancient yew tree in St Mary's Churchyard, whose *mana*, even to White, seemed beyond scientific explanation.

That was the start of my entanglement with the tree,

though who was tangling whom is no longer clear. The Selborne yew was a village landmark, but certainly not humbling or cathedral-like, or any of the other clichés that are so carelessly used about old trees. It was not even particularly big, except for its massive girth, and grew modestly on the south-west side of the church. When Hieronymus Grimm made an engraving of it for the first edition of White's book he pictured a distinctly stumpy growth, pollarded right back to the height of the surrounding cottages.

In its supreme old age it was squat and fulsome, but if one had to use an anthropomorphic simile, more Falstaffian than Buddha-like, full of knobs and quirks and a general sense of disorder. Close to, it was distinctive, with flutings and wings at the corners of the trunk. Inside, where much of the heartwood had rotted away, it was patched with a satiny sheen of lilac and green and grey, like the lustre of mother-of-pearl. And it had a seat around it.

White had written about it at some length – not in *The Natural History*, oddly, but in the accompanying *Antiquities of Selborne*, as if the tree were more one of the dead than the quick. He was aware of its great age, and admitted that it was probably 'coeval with the church'. He thought the frequent presence of yews in churchyards might be to provide shade for 'the most respectable parishioners'; or a screen from the wind; or wood for longbows; or, most likely, that they might be 'placed ... as an emblem of mortality for their funereal appearance'.

He was almost certainly wrong on all counts, and it's now reckoned that yews were not so much *memento mori* as the exact opposite, symbols, partly by virtue of their evergreen foliage, of *im*mortality. But it is the dawning realisation of their immense age which has strengthened this view, and the likelihood that their presence in churchyards has little to do with the Christian tradition.

From historical records and analysis of ring growth and evidence in the landscape it now seems certain that large numbers of churchyard yews are not so much 'coeval' with the church as vastly older than it, often pre-dating Christianity itself. Most probably they were the lode-stones round which early, possibly pagan, religious sites grew, which in their turn formed the basis for sites of Christian worship. The Selborne yew was certainly at least 1,500 years old, and radiated its antiquity. Over the centuries other writers, partly in homage to White, partly perhaps to add their own mite to the tree's history, made pilgrimages to Selborne, and put their tape-measures round its trunk. W. H. Hudson visited it. So did William Cobbett on one of his Rural Rides, and found that its girth had increased since White's day by 8 inches.

But early in 1990 the yew was blown down in a great gale. It lay south-west over the graveyard, and provoked the incumbent vicar to a declaration of awe: 'A stormy sea of twisted boughs and dark foliage covering the churchyard was pierced here and there by a white tombstone like a sinking ship.' Underneath were the remains of thirty bodies, including that of the Trumpeter, who had rallied an anti-tithe protest in the early nineteenth century. The Selborne yew was beginning to unfold its thousand-year-old memory.

After the devastation wrought by the 1987 hurricane, local people found the tumbling of Hampshire's most famous tree too hard to bear, and a rescue attempt was mounted. A team from a local agricultural college sawed off the yew's heavy top branches and winched it back into the vertical. The children from the village school, led by the vicar, linked hands round the risen bole to pray for its survival. A miracle ensued. Stirred by all the activity, an underground main burst right beneath the tree, and bathed its roots in municipal water for the next thirty-six hours. Alas, it may, ironically, have been the last straw, drowning the tree where it stood. After a few months during which it put out a few new wispy shoots, the great yew died.

But that was not before hundreds of people came to pay their last respects and carry off bits of the trimmed timber as souvenirs. Woodworkers did a brisk trade, and Selborne citizens who had lunched, courted, and proposed under its branches were able to acquire bowls and toadstools made out of the very wood that had sheltered them, and – did they think this in their hearts? – might even have incorporated some of the molecules of their breath. The church had a lute made from the wood, I went off with two raw logs, and that, you might have thought, would be the end of it.

But the yew refused to go away. The village planted a cutting taken from it just a dozen yards from the now fallow hulk, in a touching ceremony involving the youngest and oldest inhabitants. And the two logs which had spent a dozen years buried under old books and garden tools in my garage, insisted on making the journey to Norfolk with me. I gave one to David and Liza, thinking that they might have one of their beloved barn owls carved from it. But instead they took it to Mathew Warwick, an artist in wood who made them a bowl from it, using the yew to make miniature veneer inlays of White's birds.

So one weekend I carted my remaining log up to Mathew's studio and tried to articulate my own vague vision of a group of martins and swallows, or a barn owl's glare, materialising from the grain. He explained patiently why it couldn't happen quite like that. Yew is a shaky wood, full of unpredictable faults and cracks, and very difficult to turn. You cannot set out with a fixed shape in mind and expect yew calmly to submit to it. So, thinking on his feet, Mathew imagined opening the log up like an oyster, exploring its internal landscape of splits and crevices, and carving White's birds *inside* them. The habitat would determine the birds, create their niches. They would exist inside the context of a wooden retreat. This, in a way, is where White believed that some of them spent the winter, nestled down in a long sleep under the tree-roots on Selborne common. Or just possibly, though he never quite admits this, hibernating in the style of the old legend, at the bottom of the parish ponds. (Samuel Johnson invented a delectable word for their underwater clustering: 'a number of them *conglobulate* together, by flying round and round, and then all in a heap throw themselves under-water'.) Mathew took it for granted that this was what White truly believed, and saw it as a powerful myth of dwelling and belonging. I'd always thought myself that White's willingness to accept the possibility of hibernation (to the extent that he would search the local heaths and cottage lofts for sleeping birds) was less a matter of scientific open-mindedness than honest sentiment. He pined at the departure of his favourite birds – that marking of the end of summer – and in an understandable corner of his heart hoped that a few might stay for the winter in his parish. But Mathew's assumption that his belief was a kind of allegory, deliberate or not, took me aback. I'd been close literary friends with White for two decades, and it had not occurred to me to look for such deliberate metaphors inside the text. Of course, the book itself was an allegory of dwelling, a glimpse of a whole living community, 'a council of all beings'. But the evidence of White's humane and grounded rationality is so strong (and such a welcoming contrast to what had come before in natural history, and what would come after) that you need to change gear to think of him working like an imaginative novelist.

So I went back and read again the four exquisite essays that White presented to the Royal Society in 1774–5. They are still a compulsive and enchanting read. White's 'minute' watching is faultless, but the essays aren't scientific in any formal sense of the word. They're disorganised, anecdotal, affectionate. Queries are posted but not often followed up. Methodical investigation and presentation are not what he is about. Some other purpose, some alternative to the subject/object, cause/effect preoccupations of conventional science was guiding White, however subconsciously. And if you consider the circumstances and likely state of mind of the man who wrote them, the essays take on a new depth and resonance. Here was a middle-aged bachelor, confined in a remote English village, longing for intellectual company and urbane enjoyment. Writing of dwelling and migration and family responsibilities, he was contemplating not just the bird's situation, but his own, and that of all social creatures.

The house-martin essay is a story of livelihood, of a proper balance between work and play. It concentrates on the birds' busyness and good humour as builders and citizens. White spends half of the essay describing the work of these 'industrious artificers' who 'are at their labours in the long days before four in the morning ... [and who] by building only in the morning, and dedicating the rest of the day to food and amusement, give the nest sufficient time to dry and harden'. The swallows' tale is of family life, the thing the childless rector never had. He talks about how the young are brought up with 'great assiduity', of the mother's 'unwearied industry and affection' as she defends them from birds of prey, and how she feeds them on the wing: 'at a certain signal given, the dam and nestling advance, rising towards each other, and meet at an angle; the young one all the while uttering such a little quick note of gratitude and complacency'. The sand martin is the unfamiliar hirondelle, a 'fera natura ... never seen in the village ... disclaiming all domestic attachments'. It is a secret creature, 'terebrating' (White's shift into scientific jargon is quite calculated) deep burrows, raising its young in the dark, unvocal and unsociable. Its tale hints at the otherness and mystery of nature.

But the swift is White's true familiar. He celebrates its

festive assemblies and life spent almost wholly on the wing: 'If any person would watch these birds of a fine morning in May, as they are sailing round at a great height from the ground, he would see, every now and then, one drop on the back of another, and both of them sink down together for many fathoms with a loud piercing shriek. This I take to be the juncture [White was never above a good pun] when the business of generation is carrying on.' This is a story about wildness and freedom, about the chance to stretch wings that White missed, single as he was, deprived of like-minded neighbours, and cursed with coach-sickness. At the end of the swift essay he contemplates two nestlings,

> as helpless as a new-born child ... we could not but wonder when we reflected that these shiftless beings in a little more than a fortnight would be able to dash through the air almost with the inconceivable swiftness of a meteor; and perhaps, in their emigration must traverse vast continents and oceans as distant as the equator. So soon does nature advance small birds to their state of perfection; while the progressive growth of men and large quadrupeds is slow and tedious!

The notion that any of these birds spend the winter conglobulating in English villages is, of course, scientific nonsense, and it's doubtful if White 'believed' it in any ordinary sense of the word. Yet as a myth, as a truth about his feelings for the bird, it is as solid a fact as swifts dozing on the wing, way beyond our sight.

Some weeks later Mathew came over and said that he was uneasy about opening up the piece of yew, and carving something as 'finite and explicit' as an actual wooden hybernaculum. Instead, he was going to make a piece of art *about* the idea. And, by the way, they had *six* pairs of swallows nesting in their outbuildings. The piece duly arrived, a lacework of drawings and cut paper in the shape of a pair of wings. One drawing was an imagining of the birds sleeping in the log, swallows inlaid in African ebony, dreaming of southward migration. Another considered the yew log itself, and imagined its annual rings opened like the leaves of a diary, as the tree 'marks time'. 'Perhaps', Mathew wrote, the yew 'is a storehouse of the imagination, containing information and a message'. He wondered about 'the provocative nature of the possibilities that the wood encompasses', and about the possibility of passing the log on, so that different people could interpret what it contained or revealed to them.

What is one to make of this meandering dance of tree and birds and artists, in which a lump of wood seems to have behaved like a rather creative critic? Certainly that a natural science confined to the naming of parts and simplistic models of cause and effect is neither adequate nor particularly helpful in describing a world in which memory, feeling, spontaneity and a growing and necessary sense of the wholeness of things are intertwined. The nesting martins I watched at the farm had an inventiveness that can't be entirely explained away by invoking instinctive patterns of behaviour. White's account of them and the other hirondelles not only sets them in the context of a real community of many species (of which he was one), but explores the resonances between their different lives.

As for the yew, which for Mathew and me helped unlock the sensitivity of his exploration, perhaps one should simply say it has been a good listener, a kind of sounding-board whose very evident experience makes ideas bounced off it subtly refract in the mind of the observer. Yet that makes it sound too inert. An organism with such a long history of stirring things up and against-the-odds survival deserves a better billing. It would be tempting to go with the followers of New Age yew cults, and see the yew as the Tree of Life, imbued with its own enterprising spirit. I can't go down that mystical road myself, and in any case think that this would be just another subject/object view of nature, casting the yew as simply a different type of atomised individual. Maybe the gift of yews, like the 'gift of swifts' in Anne Stevenson's poem, is to help us grasp the idea of 'the mill of the world's breathing', to see that mind is a much broader entity than consciousness, and not necessarily confined in individual packages. It's a function of all life, the learned-from, responsive record that experience makes in living tissue. It can be cultural, co-operative, perhaps even communal. And maybe in keeping with our new understanding of the unity of physical life, we could try viewing mind not as *possessed* by individuals, but shared between them, as a kind of field.

*

For most of the past couple of months, the solid and reliable business behind the high jinks of the swifts and martins has been the procedures of the woodpeckers. They're everywhere. The great spotteds are still coming to the pear tree, and must be nesting close to the garden. They haunt the fens, too, revelling in the rotting alder and willow trees. On the Ling, green woodpeckers, like small dragons with their lime-velvet plumage and fiery tongues, bounce across the turf – and then bounce through the air, yelling and yaffling. Both species, maybe just two individual birds, love a particular telegraph pole just beyond the garden. They fly hard at it, in as near to beelines as their undulating flight will allow, and clamp themselves to the surface with the thump of abseiling rock-climbers. They're not feeding or probing in it, just playing sentinel, peering about, being peered at, using it maybe as a highly visible territorial look-out. Wherever I go I see one woodpecker or another, laughing, looping, levitating.

And for the third time in my life, by unspoken coincidence, they've become light-hearted billets-doux. I see a woodpecker and receive a message. Polly's thinking of me, and I of her.

Why should this idea have come spontaneously at different times to so many different people? Woodpeckers have an ancient history as harbingers. They were believed to foretell rain, the growth of crops, even the future. In the Gironde in France there is a folk-story about them as rain-seers. When God had finished creating the earth, he ordered the birds to excavate with their beaks the hollows that would become the seas and lakes. All of them complied except the woodpecker, who refused to move. God's punishment was that, as the woodpecker was unwilling to peck the earth, she (*sic*) must for evermore peck wood. And that as she would have nothing to do with making cavities for water, she must drink nothing but rain. Hence the poor bird is forever calling to the clouds 'Plui, plui, plui', and, in flight and on the ground, forever pointing to the sky so as to gather in her beak the drops which fall.

Yet other myths, from Greece to eastern Europe, portray the green woodpecker (and the black) as fertility symbols, precisely because they *do* excavate the earth, while digging for ants. What lies behind these myths? An orthodox answer is that they are both examples of sympathetic magic. The fertility myth plays on the likeness between the green woodpecker's feeding technique and ploughing; the rain-bringing legends on the similarity between the spotted woodpecker's drumming and thunder. Sympathetic magic is often simplified to the formula of 'like cures (or generates) like'; but it is really a more comprehensive (and seemingly almost universal) approach to the search for order and connectivity in nature. At its heart is the idea of analogy, the ecological, if 'un-scientific', belief that the different layers of life are not only connected, but in some way physical reflections – metaphors, if you like – of each other. Exterior likenesses are clues to inner processes and likely resonances. The shape and colour of plants reveal their powers. The mating dances of animals, if mimicked by humans, will make the animals more prolific – and maybe the dancers, too. The woodpecker thunders, and the heavens will thunder as well.

Sympathetic magic isn't some primitive stage on the upward journey towards real science. It's a different way of understanding and, its followers hope, influencing the world. It begins with observations and experiences, but then, instead of attempting to explain these by reducing them to ever smaller and more discreet parts or 'atoms', looks at them more broadly until they seem to fit into the weave of the world. Claude Lévi-Strauss called it 'the science of the concrete'. Gilbert White's sympathy with the myth of the hibernation is magical in a sense, a hopeful suspicion that birds might stay around to share his lot.

I think about the woodpeckers again, and the odd responses they stir in people. I'm far too twenty-first century to take seriously the idea that they either predict or procure rain. But they make me listen, and look up. The yaffling cry, the upward-tilted bill, the looping flights, the drumming, the flashing feathers, red and black and white – all things that caught the interest of our ancestors, catch us, too. The rhythms of the bird become the pattern in the brain. Our interpretations are more matter-of-fact now, maybe even flippant, but not entirely unsympathetic. Woodpeckers are birds of alert. They make us pay attention. They are exclamation marks, one half of a pair of crossed fingers. Of course, somewhere, someone else is also looking up, and crossing the other one.

5

Fancy Work

> A pretty little basket, probably a fruit picker for a woman or
> child. When baskets were made with fine willow for non-
> agricultural purpose, they were referred to as fancy work.
>> Caption to an exhibit in the Norfolk Museum
>> of Rural Life at Gressenhall

AND THEN, IN LATE May, after all the false starts and unfulfilled
days, summer opened, as if it had simply been waiting for the right
moment. And not just any old summer, but what was to become a
season of burnished colour and intoxicating smells that banished
elegies for days 'like they used to be' and burnt itself into eastern
England's collective memory. By a stroke of luck, I was up at dawn
on the morning it started. There was a mist hanging over the back
meadow, a thin milkiness that was hard to tell from the blowsy lace
of the last cow parsley. Then the sun came up and simply parted it,
unfolded the life of the new day from the wisps of the night. That,
it said decisively, was how it was going to be from now on.

For months the sun blazed and the wild flowers put on the
kind of show that hadn't been seen for a generation. I couldn't get
enough of it. After two summers in a dark place I was like a child
with too many birthday presents. I dithered from one wishful vision
to another. Perversely, I wanted to be in the Mediterranean and see
a bee-eater, turquoise and cinnamon against the sky. I wanted to
hear a nightingale, anywhere. I wanted to have a Greek meal in
London, outside in the street. I wanted, badly, to go back home and

walk in a beechwood. I wanted to lie all day in the garden. I wanted to have that long-awaited watery epiphany, a sail on the Broads in Polly's darting Laser. I wanted to know how to decide what to do on days that might be the only ones like them we had.

There is a simple answer to this kind of feeding frenzy, and to the paralysing excess of possibilities that freedom of choice offers us. And that, of course, is to do absolutely nothing, and let the summer happen to you. That huge moment of blooming and exuberance permeates everything. There's a legend that when the grapevines are in flower, even the most well-weathered wines show traces of effervescence, as if their remnant plant cells remembered the time they began their journey into ripeness. Maybe we fizz physically, too, as well as in our imaginations.

As it happens the choice was made for me. I had my moment of supine contemplation courtesy of nature's favourite trick, the wild card. Just a few days before Polly and I had planned to go for a break in southern Spain, I'd had a cystoscopy, an uncomfortable and undignified experience in which a TV camera is insinuated into your bladder, and a presenter gives a knowing commentary on your internal ecology. It was for nothing more serious than an overactive bladder, and produced no sinister revelations. It was what they call 'a routine investigation'. But it went wrong. Where the little biopsy specimens had been nicked out of my inflamed bladder wall, I bled. An hour after I got home all I could piss was blood. Another hour on it was just blood clots, and then nothing at all. I was blocked solid. I filled up with a diffuse and immutable pain that stripped me of the ability to fix my mind on anything else. I couldn't move. All I could do was sit very still on the lavatory seat and let it happen to me. I thought that for someone trying to adapt to a watery environment, this was a pretty black joke. I'd got bladderwrack. Just before my consciousness clouded I managed to phone for an ambulance. They were, needless to say, heroic. Manoeuvring a stretcher down an almost vertical seventeenth-century spiral staircase was a triumph for the virtues of tacking.

Nitrous oxide and deep-breathing got me red-cheeked and silly to Casualty, and to the more placid embrace of morphine. Then my bladder was reconnected to the outside world by a catheter and I felt true relief. But still I seeped. My catheter bag filled inexorably with fluid the colour of stewed plums. I was not going home yet. The holiday was off. I'd only been confined to a hospital bed once before, when I was nineteen, and viewed my enforced stay with foreboding. But it became a window on another East Anglia: the Wetland Within.

The men's urology ward was like a scene from Bruegel. Hapless men in gowns crept about in slow-motion, carrying their urine bottles like votive offerings. The bottles were made of papier mâché, and looked uncannily like the leather purses or 'skrips' worn by medieval peasants. (This is the origin of the name of shepherd's-purse, a plant that by a strange coincidence was once the herb of choice for bladder problems, especially bleeding.) This was certainly peasant-style medicine. The Filipino nurses didn't speak much English, but with a combination of body-language, pointings, extravagant sighs and a bit of pidgin, we got on well, and I negotiated my way clear of most potentially painful traumas. Polly, the great improviser, brought in preposterous foods to make up for the bland ward meals, including a yoghurt pot full of curry, which might not have been medically sensible. The beds around me filled up with men, drowning from the inside. One, laid low by a kidney stone while he was somewhere he shouldn't have been, spent the dawn in ashen-faced phone conversations trying to establish an alibi. Another, whose prostate wasn't up to the amount of beer he'd sunk on his Broads cruiser on Saturday night, had simply shut down: acute urinary retention, same as me. Feeling slightly hallucinatory, I was seized by an image of the bladder as the interface between private and public water, a barely controllable dyke that was likely to stagnate: an internal fen, with its own proliferating flora.

And still I seeped a now rather dodgy-looking plum wine. On the fourth day, it was decided that I was thoroughly stagnant with undissolved clots, perilously close, no doubt, to peat. I needed sluicing. So I was given a bladder wash, an ordeal by water in comparison with which cystoscopy was a soothing massage. The sharp-nozzled injection device that was repeatedly shoved and emptied into my bladder (and then sucked out like a stirrup-pump) might just as well have been the staff-nurse's whole hand, so overwhelming was the sense of intrusion. My romantic yearnings to live more intimately with the outside world were being thrown back in my face. The novelist Joyce Carol Oates had had a similar sensation. Whenever she was laid out by attacks of the frenzied heartbeats known as paroxysmal tachycardia, she experienced a compelling sense of the aggressive indifference of nature: 'When you discover yourself lying on the ground, limp and unresisting, head in the dirt, and helpless, the earth seems to shift forward as a presence; hard, emphatic, not mere surface but a genuine force – there is no other word for it but *presence* . . . The interior is invaded by the exterior. The outside wants to come in, and only the self's fragile membrane prevents it.' Lying on that side-room table, I felt that the outside world *had* come in, and breached every one of my fragile, protective membranes.

But it seemed to do the trick. I filled two skrips with the clearest golden liquid, and was discharged that same afternoon. But I was feeling disorientated and invaded. My body seemed like an unreliable stranger, no longer part of me. It was an unnervingly familiar sensation. I saw my hard-wired escape route looming, that senseless retreat away from the world, and knew that this was the testing time.

Polly, thank goodness, had the nous to take me straight out into a wild spot, and meet that 'hard, emphatic' presence on more even terms. We went to the fen at Strumpshaw on the western edge of the Broads. It was a mild, hazy afternoon. I was shaky on my feet, but said the mantra to myself. Listen. Look up. It wasn't easy, this time for the best of reasons. The fen flowers were beginning to unfurl at my feet – ragged robin, forget-me-not, yellow iris, a *double* cuckooflower, the first I had ever seen in the wild. On a thin blade of grass a female orange-tip butterfly, her almost translucent wings dappled with small clouds of grey, shrank to a raindrop in a gust of wind. I felt myself pull in to the wind, too, but catching it, not cowering. We meandered on. I introduced Polly to the pepperiness of lady's-smock, and she made me a water-mint sandwich between two of the oatcakes that are always in her bag. In the willow-scrub at the edge of the fen, a Cetti's warbler exploded into song. I heard it without the help of Auric – and heard reed and willow warblers, too. At the edge of my vision the harriers patrolled the reed-swamps, coasting on contours of the air encoded by the intricate textures of the marsh beneath. I felt as if I were unfurling myself, reconnecting with the grain of things. We walked along crystal-clear dykes fringed with wild blackcurrant bushes, and meadow rue and the elvish shoots of cowbane. As we were turning for home, we came upon a cottage lying by itself at the back of the fen. It had a long, classic herbaceous border, that sloped down from the house to the edge of the swamp, full of delphiniums and stocks and lupins. And by the path, at the edge of this fabulous display of cultivation, was a sign that could have crept up from the common fen, an indication of a generously permeable membrane with the wild: 'If you are watching swallowtail butterflies, please feel free to follow them up our border.'

*

I was slowly getting in touch with the fens. I'd still go home from my walks drenched to the ankles, failing yet again to spot where spongy ground gave way to quaking swamp. But I was beginning to find my way around, to read the subtle layers and textures of the vegetation. They made, in Weston Fen especially, quite a show. People often use the metaphor of a tapestry for these kinds of display, but that makes them sound too symmetrical. What was going on here was a vegetational plot, an ingeniously crafted scheme for the occupation of land, full of flexibility and give-and-take, but with each participant playing its best role.

It begins in the open water of the pools, with the bright yellow bobs of bladderwort, held on stems above the water and swaying in unison in the breeze. Bladderwort's leaves are entirely submerged, and covered with tiny sacs that sieve and trap and eventually digest minute water organisms. It is on the outlaw edges of the fen vegetation, a diminutive vegetable whale. The shallower pools are rimmed with mosses, absorbent sponges that, before they rot away to peat, become a temporary fen surface themselves. Sundews, another insectivorous plant, once grew on these sodden mosses, but it has gone now, or become too scarce to find. But there is a third insect-eater at Weston, the violet-flowered butterwort, sometimes known as 'the starfish plant', from its rosette of sticky yellow-green leaves, slapped out flat on the peat. It haunts slightly more solid ground, where the water from the pools and springs seeps in runnels through the short turf. With it is often marsh valerian, its delicate pink umbels scented with just a hint of the cloying vanilla of its larger cousin. And creeping through them all, from the edges of the pools right into the thickest reed-stands, is water mint, scenting every step you take.

But centre stage are the orchids, a teasing, shape-shifting

swarm of hybrids between common spotted orchid and up to three species of marsh-orchid. They are no easier to pin down botanically than the *Orchis* of the Cevennes, and maybe it is pointless to try. This freckle of magenta and pink and white, of toadish heads and elegant tapers, may be one protean species, or a thousand. The species that were split up when botanists first began systematically naming our orchids may now be merging, or just trying out some new experimental crosses.

But one species is not in doubt. It isn't made fuzzy by hybridisation or retiring habits. In July the marsh helleborine comes into flower, and turns the back of Weston Fen into an outpost of the tropics. This, for me, is the most glamorous of all our orchids, the one that seems closest kin to the aerial extravaganzas of the rain-forest. When I first found them here, I could scarcely credit the numbers. I had only ever seen them once before, a small colony jostled by reeds in a damp sand-dune hollow up in the north of England. But at Weston there are thousands of plants. Each stem has between ten and twenty blooms, carried on stalks arched like swans' necks. Under the starched, three-pointed cowl, the flower's lower lip is pure white, frilled and crimped at the edges, like a fop's handkerchief. They are handsome and totally gratuitous. In the gipsy clothing of the fen, they are Saturday-night Best. But what is also striking is the sense they give of habitation. They have poise, hovering with the delicacy of a dipped toe just an inch or two above the water. They cling to the hummocks as if these were life-rafts.

All these shorter plants grow mainly on those parts of the fen which are cut annually, so that the surrounding vegetation is also low early in the summer. In the less kempt areas, in some of the larger areas of water and along the dyke tops and edges, the taller plants hold sway. This is the domain of the vertical, of bulrush, reed, the blades of yellow iris, of meadow rue and the claret-tinged florets of hemp-agrimony, and overwhelmingly of the sprays of meadowsweet. Meadowsweet is the froth of the fen, a plant whose scent – honey and marzipan in the flower, something cleaner and almost antiseptic in the leaves (it was called courtship-and-matrimony from this contrast) – is the counterpoint to the sharp tones of water mint. Its tones weave the whole fen together: the cucumber-ness of high summer, the astringency of clear water, the hint of mown hay.

In open patches among the reeds and sedge-lots there are also humps of tussock-sedge. This is the densest and most aspiring that fen vegetation becomes without actually turning into woodland. Tussocks are sedge-trees, their root-masses and dead fronds accumulating beneath them so that they rise up out of the swamp, sometimes as much as 4 or 5 feet. In some exceptionally wet and unmanaged areas of the Broads tussocks become another, aerial surface, a base in which alder and willow seedlings can take root. When the trees become large and heavy enough, they fall over and take the tussocks with them. The water partially closes round the remains, and the whole cycle of oscillation between wood and water begins again. I doubt that this would be allowed to happen at any of our valley fens, so protective a view is taken of their pure fennishness. But it would only be a transient stage, and for a while at least the slender sheavings of the ancestral fen would return.

The style of the fen plants, an almost universal upward-reaching slenderness that extends way beyond the grass family, is something of a puzzle. Why, in soil so rich it is still pilfered by gardeners, isn't the ground monopolised instead by some light-hogging bully, some pushy, parasol-leaved Greater Fen Dock? (The real great water dock is the most restrained and finnicky of plants.) It is as if they were designed for multi-cultural living. The assumptions of crude Darwinism and the selfish-gene theory are that lineages of plants (and, by proxy, the species which they represent) are perpetually striving to expand their territories and out-compete their neighbours, in order to maximise their off-spring's chances of success. But what happens in the real world doesn't much resemble a simplistic race towards winner-takes-all homogeneity. In the fen, there is a general movement towards the development of woodland, unless this is prevented by flooding, grazing, or deliberate cutting-back, and for a while this shading will mean a reduction in diversity. But against this there is a corresponding, intrinsic drive towards variety, flexibility, subtle forms of symbiosis and partnership. At any opportunity, any slight break in the canopy, the vegetation and all the life that depends on it will proliferate and diversify – over the surface, down through the layers of soil and damp, and through time, too. The natural long-term tendency of any ecosystem is to become progressively more plural, complex, sociable. Plants live in the fen in such close and dense harness that it seems more than a matter of passive tolerance, of each species holding its ground simply because it can't encroach any further.

Is intimate mingling some kind of optimum state, where different species benefit in each other's company, enjoying maybe the modifications of the soil produced by their neighbours, and profiting from the specialised chemicals secreted by their roots to tie in symbiotic fungi or deter predators? Meadowsweet, for instance (dubbed *Spiraea ulmaria* in the early twentieth century), lent its name to aspirin, acetylsalicylic acid, the world's most useful drug. This occurs in some concentration in the plant, as it does in willow, and, to a lesser degree, many other species, where it acts as an anti-stress chemical. Might meadowsweet's aspirin be leached into the soil in the fen, to other species' happenstance advantage – a kind of natural companion planting? And is this kind of symbiosis the reason that newly arrived exotics can sometimes be disrespect-fully colonial, not because they lack 'natural predators' (what on earth consumes the impeccably native meadowsweet?) but because they are not yet part of that anciently evolved network of chemical reciprocity?

The fens, in Bourdieu's word, are a habitus, a field of play and natural possibility. They are the nodal points in the valley's water system, which stretches from the sugar-beet stands where the house martins gathered their mud, through the roadside grups and gutters, into the ditches and streamlets with their reed-stands and willow-herb thickets, connecting beneath the surface with the moats and ponds and peat-pits on the springlines, and eventually with the rivers themselves, the whole system forming a conduit for everything from moorhens to meadowsweet roots. And walking in the fens this summer, I've felt, in the most flattered way possible, water-shaped myself, caught up in the current. I'm momentarily one of the company. I ferry seeds, stuck to my shoes. I make brief openings in the reed canopy every time I peer across at a pool. Whenever I step onto the peat – even in this baking season – tiny efflorescences of moisture spread round my feet, and I have the feeling that yards, maybe miles, further on, I'm squeezing water out onto some slumbering aquatic growths. The wind is shaping things, too. It's bunching up the sweet-grass, entangling it, making nets that will play their part in gathering peat dust and seeds into a brief anchorage. It's carrying scents, mint and pollen, and the burnt-mushroom smell of peat itself. Near my feet the first froglets are moving like mudfish, scrambling out of the slime. Dragon-flies shoot past me at eye level. They move with the speed of animated cartoons, seeming to be first in one spot, then another, without ever having travelled through the space in between. I can't tell if they make any sound with their wings, but they pass so abruptly, make such impossible turns and jolting stops, that they seem to make some kind of brittle snap in the air.

The fen, in these summer moments, seems more than a

habitat; it's a membrane, pulsing with interconnected life, busy with communications. Joyce Carol Oates, felled to the ground, viewed the 'self's ... membrane' as something fragile and protective, a barrier to the 'hard, emphatic' world of nature beyond. Thoreau took a similar view of, as it were, 'nature's membrane'. But, in contrast to Oates, he experienced a kind of ecstasy when it was breached. After his epic climb into the desolate wilderness on top of Mount Ktaadn, he was like a prostrated pilgrim, or a flagellant: 'Talk of mysteries! Think of our life in nature, – daily to be shown matter, to come into contact with it, – rocks, trees, wind on our cheeks! The *solid* earth! The *actual* world! The *common sense! Contact! Contact!*' The fen's membrane is altogether more magnanimous. Not fragile particularly, not defensive, certainly not mysteriously overwhelming, but tactile, inclusive, accommodating. For all the antiquity of the peat, life at the water's edge is nimble and adaptive, about living in the present, going with the flow.

In the 1830s Clare was living away from home in Northborough, a few miles east of Helpston on the edge of the Cambridgeshire fens. He'd been found a house there by his friends and patrons, in a well-meaning attempt to ease the pressures caused by his growing family and declining health. But the move from his familiar home ground may well have been the last straw, an overwhelming final alienation to add to the losses in his life. He had been shifted, as he had once put it, 'out of his knowledge', and it was at Northborough that he wrote his poem of exile, 'The Flitting':

> Strange scenes mere shadows are to me
> Vague unpersonifying things ...
> Here every tree is strange to me
> All foreign things where e'er I go

> There's none where boyhood made a swee
> Or clambered up to rob a crow.

The point was not that Northborough's landscape was dramatically different from Helpston, but that it wasn't, in its specifics and individuals, a *known* place. But Clare gradually found consolations – a tuft of the same shepherd's purse that grew in his old garden, a wisp of woodbine round the gate. And the snipe, an old friend ('often seen with us in summer', he wrote of the bird in a journal note about six years previously, after finding one nesting on the common), and now another solitary in a flooded landscape, a new ally. His great fen poem, also written at Northborough, is titled 'To the Snipe'. It is dedicated and addressed to the bird, out of respect and sympathy. In it, the fen is pictured from the snipe's point of view, a wilderness towering above the quagmire:

> the clump . . .
> Of huge flag forrest that thy haunts invest
> Or some old sallow stump
> Thriving on seams
> That tiney islands swell
> Just hilling from the mud & rancid streams
> Suiting thy nature well.

Clare's ramping dialect wonderfully catches the swampiness, the strangeness and privacy of their shared exile:

> The little sinky foss
> Streaking the moores whence spa-red . . . water spews
> From pudges fringed with moss.

Like Clare, the snipe was seeking refuge from 'man's dreaded sight', from 'the freebooters' that would destroy and 'betray' its 'mystic nest'. It was an embattled fellow creature, an inspiration in the struggles of dwelling. 'Thine teaches me', Clare ends in gratitude, 'Right feelings to employ':

> I see the sky
> Smile on the meanest spot
> Giving to all that creep or walk or flye
> A calm and cordial lot.

*

While the fen baked in the sun, the whole countryside went *en fête*. In our market town, Diss, half the population went into hot-weather moult and partied by the mere. So did the local swifts, whose flight games over the water were little short of pyrotechnic self-indulgence. In the hottest weeks I watched them deliberately slaloming through telephone wires, and hurtling in criss-crossing ribbons between the back-to-backs. Out in the villages where there were big nesting colonies, there were hullabaloos of swifts. The excitable teenage birds were pushing and threshing near the entrances to their elders' nests, trying to get in, and then taking on the traffic, flying in low mad chases across the bonnets (though it occurred to me that I've never once seen a swift road-kill).

Just walking about felt like taking part in some summer festival. There was none of that common summer feeling of *tristesse*, the sense that things are over, *spent*. This summer felt as if it would go on for ever. On the Ling the scenes were as far removed from 'the tragedy of the commons' as you could imagine. This was the comedy of the commons. Clare had written about 'commons wild and gay' but this was outrageous. Garish mixtures of purple and lilac and yellow spilled over what remained of the parched and rabbit-raddled grass. Rosebay willowherb clashed gloriously with the first heather blooms. Anciently naturalised escapees – soapwort from some long-gone riverside laundry, and everlasting pea from the Mediterranean – bloomed next to post-glacial indigens like meadow rue. A small grove of Japanese roses came into flower above their gnawed stems and raised the possibility of the emergence of their rare hybrid with the dog-rose, *Rosa x paulii*, an exotic and mysterious rambler with long white dandified petals. And as a testament to the geological oddity and general contrariness of the Ling, the meadowsweet of the fens grew cheek by cheek with its closest cousin, the dropwort of the dry chalk.

It was turning into a carnival of all beings. Bee orchids sprang up on roadside verges and back lawns. On Wortham's moated Long Green, a hybrid swarm of marsh and spotted orchids appeared – 'very plentiful', as Clare might have noted, on a spot near the Road to Spears Hill just past the War Memorial. Everywhere cottagers were putting their bounty of surplus flowers and vegetables on old tables for sale: a single cabbage, 50p; three bunches of pinks in vase. On some of the greens, the travellers' fancy horses came out in numbers, and I began to understand what, in the absence of any obvious physical role, they were for. They were, a little like Arab camels, part status symbols, part currency. Big black spots and long blond fringes were the most highly valued ornamentations, the ones that might be cashed in, or traded.

Then it was the insects' turn. The house walls became so speckled with swirling hieroglyphics and living graffiti that it felt like living in a Joan Miró painting. 'Miró's signs and symbols', a critic once wrote, 'hover in the ambiguous but shallow space created by his staining' – which sounded a pretty good description of Ian's newly painted walls most evenings. First came the flying ants, mundanely, but in profusion. They poured out of every crack between floor and walls. Back in the Chilterns I had got used to the more orderly flight displays of the black ants in the lawn, which occurred simultaneously, on the same high summer day, over a wide area. The spiralling mating-flights of the winged males and females, pouring upwards in iridescent plumes, were so spectacular that 'ant-rising day' was an annual feast-day in our family calendar. But here every day was turning into ant-rising day, a sign, I imagine, of an occupation as ancient as the house itself.

As the temperatures continued to creep up, humming-bird hawk-moths arrived from the continent, and spent breakfasts darting between the honeysuckle and the muesli bowls. One evening I watched four ghost-swift moths perform their mesmerising 'lekking' tarantella, in which the males dance and hover around the females, rising and falling with wings vibrating so quickly that they look like small balls of mist. Will-o'-the-wisps have vanished from the fens, but this was the next best thing. When it was dark, the windows framed intricate, heraldic displays of moths: swallow-tail moths, small magpie-moths, the feathery arms of plume-moths.

It was watching these animals going about their duskings that, in more than one sense, began to bring home the insect world for me. It was easy to make fun and fables out of them, to reduce them to comic props in scenarios of the heatwave. But when I watched them for long periods, they weren't behaving like trivial cartoons at all. The senseless moth of popular mythology, consuming itself in a flame, is a creature disorientated by enclosure as much as light. Outside, their vigils against lamps are calm and meditative. Like more complicated animals they have a sense of place and territory. When the crickets came in, it was clear that they were as intent on being house-guests as the hirondelles. Gilbert White loved crickets above all other insects. The first piece of imaginative prose to break into his rather pedestrian gardening journals is a eulogy to the crickets in a field near his house. The female was 'dusky'; the male was 'a black shining Colour, with a golden stripe across its shoulder like that of the Humble bee'. He would, he felt, be 'glad to have them encrease on account of their pleasing summer sound'. They 'had express'd distress at being taken out of their knowledge', and he worked out a way of encouraging them more gently out of their holes: 'a pliant stalk of grass, gently insinuated into the caverns, will probe their windings to the bottom, and quickly bring out the inhabitant; and thus the humane enquirer may gratify his curiosity without injuring the object of it'.

Our green bush-crickets arrived just after the green tortix-moths, and stalked everywhere (including the beds) on their prodigiously cranked-up legs. But the dark bush-cricket turned up in the sewing box, exploring chunks of magnet. That same evening I found him squatting on a dimmer-switch. He was, I think, a hippy cricket, fond of good vibrations. My field-guide said his night-time song was a staccato 'psst-psst'. Something was going on here.

This unscheduled intimacy with insects began to have an effect on me. I imagined the house not as an invaded (or invading) colony, but as a kind of carapace, a complex living shell generated by all its occupants. Among these, of course, were the cats. Blackie especially joined in the spirit of things in elaborate new play. She'd taken to a trick that I've encountered in cats before – throwing herself in front of me as I was walking, inviting me to tickle her vunerable underparts, then writhing free, darting up and doing it all again a few yards further on. This was more than the soliciting of physical pleasure; she was deliberately goading me, enjoying my own reactions, curious about how far I would go. I felt I was the Pavlovian guinea-pig, not she. Play in adult cats, in which they more or less regress to kittenhood, is a fascinating business, not at all explicable as a training for the grown-up responsibilities of hunting. It is often assumed to be a luxury, available only because domestication gives cats 'free time', and not to occur in genuinely wild animals. But it does, throughout the natural world, in displays which clearly show a relishing of sensual experience and interaction with other creatures (and sometimes other species).

Cats take an evident and unqualified pleasure in doing mundane things, in repetition, and in satisfying their fabled curiosity, that goes far beyond the limits of biological usefulness. My Chiltern cat Pip (so called because he began life looking more like a large-eared pipistrelle bat than a kitten) often played experimentally. He frequently crossed the species barrier in his explorations (and with dragon-flies, alas, often simply incorporated it) and once had an encounter with a deer that made my own face-to-face meeting seem a casual and timid affair. Pip had always been a great sniffer. He would, each morning, lie on the patio with his eyes wide open but seemingly looking at nothing, then go for his morning read, beginning always at the same rose twig. He examined it minutely, and it wasn't hard to imagine him in spectacles; but he was of course smelling the night's news. He ran his nose meticulously along the twig, deciphering his own and others' scents: cats, humans, badgers and foxes perhaps, specks of bird droppings, maybe the growing twig itself, one day older than when he last read it. One day, he picked up the scent, and doubtless the sound too, of a monkjack deer. I had glimpsed it a few hours before, chewing its way meditatively along the herbaceous borders, giving an overdue prune to the shrub roses. Once Pip had spotted it he slithered up to it on his stomach. They both stood stock-still and tentatively rubbed noses – then sprang apart in surprise at their temerity. Then they tried it again, and this time, Pip let the monkjack wash his face. The word of this visitation somehow got around, and in a matter of minutes the entire neighbourhood cat gang arrived, stalked up to the extra-terrestrial and took it in turns to gaze in astonishment, exchange sniffings, and shake a paw. The deer seemed unperturbed and went on chewing roses between the odd tongue-wave and nod in return. Only when the cats' attention became too intense did it occasionally shape up for a mock butt, or leap backwards, which paradoxically seemed to have an even more chastening effect. It stayed in the garden for a couple of days, and rested up in a kind of form under our yew tree. And while it ruminated on rose cud, up to half a dozen cats would be squatting at various distances from it, gazing at it with tireless fascination and seeming to become just as relaxed themselves. It was, for a while, like one of those Renaissance paintings of Eden before the Fall.

The heat seemed to be underlining the sensuousness of the world, just as the winter cold deadened it. The valley began to be filled with unaccustomed smells, as if it had been shifted south about a thousand miles. The fugitive scent of wild roses hung around the hedges. Gorse bushes hummed like coconut groves. Peat dust became a kind of primordial vapour, like the miasma that had blown out of the house's woodwork in October. And, there being no discrimination in these things, the still evening air was full too of the stench of fungicides and growth-retardants and waftings from the feather factory in Diss.

There were new — and remembered — sounds too. Tar-bubbles in the lane burst with the softest of pops, like smacked lips, a noise I hadn't heard since I was a child. On the Ling, desiccated stag's-horn lichens crunched underfoot. One sweltering June evening Polly and I went off west to listen to nightjars, whose churring song is the epitome of hot summer dusks. We'd failed at what had once been one of their favourite haunts, Knettishall Heath, where sweeps of short heather and grasses were broken by a scattering of pines. It seemed a perfect site, and we surprised a long-eared owl in the heather. But there were no nightjars.

So, a month later we drove out to Breckland, where, in a curious turn of events, the birds are deserting the heathland and colonising the huge areas of clear-fell in the forestry plantations. In a way this may be the awakening of some deep encoded memory in them, as for most of its history Breckland was a place of brief, cultivated clearings in the wildwood. We had a picnic by the river in Santon Downham, named from having once been half buried in a sandstorm in 1668, and watched the swifts flying over the forestry-workers' houses. But it wasn't a sensible indulgence. We got behind schedule. Nightjars normally start singing about three-quarters of an hour after sunset, and we had about ten minutes in which to find a spot, if we wanted to enjoy the whole show. We drove into Thetford Forest for about a mile, and stopped at the first likely clearing, about 20 acres of clear-fell surrounded by young saplings and a ring of older pines. We stumbled off through the shoulder-high bracken, thinking it would soon thin out. It didn't. A quarter of an hour later we were still struggling through this ferny jungle, barely able to see each other, let alone the world outside. But it was too late to turn back. The ritual overture of the midsummer dusk, familiar to all nightjar *aficionados*, had already begun.

The light fades, and our not-yet-adjusted eyes are tricked by those brief, luminous mirages known as phosphenes. The woodcock enters left, stocky and log-like, flying laboriously round its territorial stage in the display known as roding. It grunts, an inward, reptilian noise that seems barely a communication at all. A few hundred yards away a roe deer barks. It's stifling in the bracken depths, and the moths feel as if they are flying through our hair. The sound begins imperceptibly but unmistakably, like the first few turns of a distant engine. Then it swells, an incessant, throbbing ratcheting. The fern owl is at least a 100 yards away, but its churring seems to fill the air. We blunder through the bracken again, trying to get closer. The noise is narcotic, it shuts out every other sound. It rises and falls as the bird turns its head from side to side and fills and empties its lungs. We get close to the tree it's singing from, and abruptly it stops. The silence is sudden and shocking, as if a plug had been pulled out. Then it's flying, glancing out of the tree with its wings held up, bouncing and floating above the bracken. Another bird is following it, and we can see the white wing-and-tail spots of the male. They vanish, somewhere among the young pines. We wait again. It's almost completely dark now. Far away to the north, another nightjar begins, a thin staccato reeling that seems like a sound in one's own head. Then suddenly the first bird is back, only a few trees in front of us. Its churring is stupendous, archaic. It rises and resonates in the close air between the pines, pouring out for two, three minutes. What is it for? With a song like the nightingale's, intricate and improvised and unquestionably 'musical', it's easy to believe that the bird experiences interest and pleasure in its own performance. But the nightjar's primeval growl seems something alien, fabulously ancient, the rumblings of a mechanism not quite come to organic form yet already hugely excessive and thrillingly florid. 'I'm here, where are you', it says in the statement of identity common to all birdsong. But it carries half a mile even to inadequate human ears. What else is it saying? 'Will you please keep your distance', to the nightjar to the north? 'I am still on call', to its mate? That this evening it's frisky, hungry, suspicious? But it may be wrong to look for precise messages. Even some scientists are coming round to the view that, in part, birdsong may be a purely emotional outburst, an outpouring of sheer aliveness, and without referential meaning. In which case it is genuinely closer to music than language.

And it is an individual expression as well as a social communication. Close listeners say they can distinguish the churrs of individual birds. Can other animals too? What can this overwhelming avalanche of sound mean to the roe deer, with its exquisite hearing? To the woodcock who barely deals in sound at all? Are roosting warblers disturbed by it? Does it form part of an ensemble of music through which all the creatures of this place make maps of their intimate territory? Lewis Thomas once wrote of the earth's 'grand canonical ensemble':

> The individual parts played by other instrumentalists – crickets or earthworms, for instance – may not have the sound of music by themselves, but we hear them out of context. If we could listen to them all at once, fully orchestrated, in their immense ensemble, we might become aware of the counterpoint, the balance of tones and timbres and harmonics, the sonorities ... the combined sound might lift us off out feet.

I got my sail in the Broads eventually, in a comely craft known as a White Boat. Needless to say it didn't go entirely to plan. We'd chosen the one summer's day when a gale was blowing. Polly and her sister Clare took two whole hours to get the boat away from its moorings, as it was repeatedly dashed back by the wind. All too aware of my total ignorance of boating lore, I kept quiet and out of the way, and thought of what G. Christopher Davies had once written about ladies on boats. Davies had been almost single-handedly responsible for popularising the Broads during the late nineteenth century, to the extent that, in later editions of his *Handbook to the Rivers and Broads of Norfolk and Suffolk*, he had to devise codes of etiquette to guide the hordes of tourists he'd attracted. 'Ladies,' he'd pleaded, 'please don't gather armfuls of flowers, berries, and grasses which, when faded, you leave in the boat or yacht for the unfortunate skipper to clear up.

Don't play the piano in season and out of season (the reedbird's song is sweeter on the Broads); and don't turn out before eight o'clock in the morning when other yachts are near.' If only he could have seen Polly and Clare, arms full of biting and thrashing ropes, and looking like guerrilla fighters as they struggled to set the boat free. But we got away in the end, and raced about as if we were a foot above the water. I was allowed to take the helm, and for five whole minutes managed to avoid capsizing or – just – making a fool of myself. I tried to hold the sheets and the tiller as lightly as I could, to feel the geometry of that equation between the tilted sail and water-pressed rudder. I wished I had three eyes, one for the water, one for the mast-head streamer that showed which way the wind was blowing, and one for what was happening over the reeds. There was a wild-fire smouldering through them, and marsh harriers were patrolling the edges of the smoke plume, looking for refugees. One was moving in the same direction as us, and in a moment of mad self-confidence, I tried to line up the sail with the dihedral of its wings, which were carrying it – as our sail was us – in that seeming miracle of moving against the wind. Then it did a 90-degree tack on a pin-head, which would have had us over if I'd tried to copy it. I know my limits, even though they're changing. But a sentence formed in my mind, about as deeply philosophical as a motto in a cracker, but seeming to catch something of what I had learned this last year. 'In the wind', it went, 'you cannot move between two places in a straight line.'

I think about how much my own perceptions have changed. Two years ago I had a Russian orthodox bass singing in my ear. Now it's the reeling of nightjars and the confidential chatterings of reed warblers. In the torrent of sensations brought by the heatwave, the electric flicker of bats in the dusk, the scent of parched grass, the velvet sheen of bee-orchids, my night vision and my sense of smell have both sharpened up. And, most crucially, my attention. What kind of attention do the other organisms in the system have? Is it discriminating and functional, wiping out all irrelevant sensory 'white noise'? Do the ghost swift moths 'notice' the smell of baked peat as well as their food-plants? Does a bumble-bee ever register a bee-orchid, and fulfil the relationship that this plant was theoretically designed for? What do screaming swifts make of the rings of mobile phones and the high whines of the motor-bikes they spent their days flying over?

So many of the transactions between different organisms seem almost incidental, wildly gratuitous. Quite late in the summer I was idling by the big house lake, when a little egret drifted in, pale and virginal. It flew more like a barn-owl than a heron. I doubt that the resident birds had ever seen anything like it before, and everything from moorhens to greylags went for it, a screeching, pecking mob of vigilantes that harried it across the water and into a tree. But when I went back the following day they'd calmed down. I spotted the egret, as weightless as a shawl, in a group of about 500 lapwings, on one of their take-it-or-leave-it fly-rounds. It followed their every move, matching their wingbeats, aping their jinking 'swoons' before coming into land. Was it seeking reassurance, or company, or just an outing in the sun? At times it seems as if the whole company of nature, ourselves included, is simply at play.

In 1974, the last full year of the Vietnam War, an American English professor called Joseph Meeker wrote a book called *The Comedy of Survival*, which, oddly for the time, combined literary criticism with a look at animal behaviour. Its thesis is the value of what the author calls 'the comic way' as a stance for perceiving the world and a strategy for living. He is at pains to point out that comedy is not necessarily humorous, but that it contrasts starkly with tragedy, with its devotion to abstract moralities, and struggles for power, and the inevitability of disaster. The workings of the natural world, are, in Meeker's terms, essentially comic. They're about durability, survival, reconciliation. Evolution itself

> proceeds as an unscrupulous, opportunistic comedy, the object of which appears to be the proliferation and preservation of as many life forms as possible. Successful participants in it are those who live and reproduce even when times are hard and dangerous, not those who are best able to destroy enemies or competitors. Its ground rules for participants, including people, are those that also govern literary comedy; organisms must adapt themselves to their circumstances in every possible way, must studiously avoid all-or-nothing choices, must seek alternatives to death, must accept and revel in maximum diversity, must accommodate themselves to the accidental limitations of birth and environment, and must prefer cooperation to competition, yet compete successfully when necessary . . . Comedy is a strategy for living that contains ecological wisdom, and it may be one of our best guides as we try to retain a place for ourselves among other animals that live according to the comic way.

The ultimate expression of the comic way is play, an almost universal phenomenon among more complex animals (and which includes what humans call art), and one which, in its exuberant purposelessness, seems close to the heart of the whole business of life. Play is the opposite of Management by Objectives, the current creed which rigidly screens out spontaneity, imagination and surprise as parts of the creative process. A play ethic with rules would be a contradiction. But Meeker suggests a Playbill of Right for all creatures – subject of course to tacking, negotiation and daily revision:

> All players are equal, or can be made so
> Boundaries are well observed by crossing them
> Novelty is more fun than repetition
> Rules are negotiable from moment to moment
> Risk in pursuit of play is worth it
> The best play is beautiful and elegant
> The purpose of playing is to play, nothing else.

6

The Wild Card

'In wildness is the preservation of the world.'
Henry Thoreau, *Walking and the Wild*, 1851

'Wild thing, you make my heart sing.'
Reg Presley, of the Troggs, 1965

The first rains came early in September. They were light and brief, but changed things. Acting again from agendas quite secret to us, the swallows and martins decided against final broods, and left *en masse*, almost overnight. So did the travellers and their horses, and the commons seemed momentarily emptier. But on Fair Green, now free from grazing, there was a strange flowering which was hard to credit as entirely spontaneous. A large fairy-ring of grass had somehow organised the flowers around it into a bedding display: inside the circle was pure white yarrow; outside a candy-floss corona of pink yarrow and late yellow lady's-bedstraw.

A hundred and fifty years earlier Fair Green had been the site of one of the most colourful fairs in East Anglia: a thousand sheep, 'droves of ponies', wrestlers, an 'Exhibition of Mechanical Figures', raw herrings, seaside rock, and prodigious drinking. It was so rowdy it was shut down by the Home Secretary in 1872. Now, its descendant in spirit, if not in disorder, is the autumn Greenpeace Fair, the climax of the season for Alternative Waveney. We all went along, and happening by chance on the day after the great exodus of the migrants, it had the feel of a Harvest Home.

There was an East Anglian reggae band, and some very passable flamenco dancing. On the sound-stage the power came from a bicycle-driven dynamo, manned by unflagging volunteers, though lacking the majesty of the giant six-seater of the Green Gatherings. The young from the pied economy sold home-made jewellery and Taiwan-made shirts, and had what sometimes looked like their entire worldly possessions laid out on the grass: old towel rails, rusting garden tools, torn dresses, back numbers of the *Ecologist* interleaved with *Asian Babes*. This was the counter-culture's car-boot sale. We ate a Thai curry from the kitchen tent, bought a basket of bright little Robin pears, Norfolk's very own variety, from a donnish lady who had them growing in her garden. Polly was called on to exercise some old skills in reviving a man who'd fallen dead drunk while dancing to the Suffolk Samba and I got into a row with some hucksters who were selling electronic devices to fry insects, stoutly pleading the parity of the wasp with the whale in the council of all beings. 'Get a life,' they snorted back. It was all delightfully mad, and, nibbling pears and listening chokily to the absurdly poignant noise of the bike-driven dynamo, I wondered why life couldn't always be like this. I was only half fanciful: I was toying with the idea of a stall myself next year, peddling remaindered books in return for plates of cous-cous and recycled shirts.

The damp passed quickly away, and another Indian summer set in. The fen hedges were festooned with hops. The wild fruits were prodigious and, with a predictable display of national masochism, were interpreted as signs of a hard winter to come: we were going to have to *pay* for that summer. Less morbidly the crop was attributed to the drought, as stressed trees threw their last energies into producing offspring. (But as my friend Sue Clifford once commented, it would make just as much biological sense for plants to make their most plentiful fruiting when they were extremely unstressed, and enjoying perfect growing conditions.)

I got into foraging again. I'd started back in the spring, when the first reed-shoots were poking through. A half-memory from an American wild food tract (a false memory as it happened; I should have been after the sap which oozes from the broken stems) made me prise one open and nibble the sappy white pith. It was startling, a fresh, lemon-zest and sugar bite, like sweet vernal-grass writ large. It was a perfect subject for what the 1930s fruit gourmet Edward Bunyard called 'ambulant consumption': too fiddly to tempt you into irresponsible over-picking, but wonderful as an occasional quirky treat. I was soon back munching anything, as I did thirty years before. I tried very young hop shoots, that great bounty of the fen, first raw (a tad bristly), then cooked in omelettes (nutty but stringy). I found a stand of winter-cress on some waste ground by a filling station and chewed the leaves. They were past it, leathery and bitter, so I tried the yellow flower-buds instead, and found them like peppery broccoli florets. From there it was just a short step to poaching oilseed rape buds from the edges of arable fields (a justifiable tithe, it seemed to me, in compensation for that malodorous crop), and then trying every possible non-toxic flower-spike from red dead-nettle to comfrey. By midsummer, the best thing was picking sorrel on the greens, especially in the evenings, when the low sun was shining through the spangled orange clouds made by the flower-heads. We made it into an ethereal green soup with yoghurt. And in late summer, coming back from a failed expedition to find Norfolk cranberries in the boggy depths of Cranberry Rough, I found a feral pear tree in a hedge. It was huge – 40 feet tall and 4 in girth, and underneath were russet pools of the little Robin pears we'd bought at the fair. I gathered 10 pounds off the verge – an entirely unexpected and unbidden crop.

But in this spectacular autumn weather, it was now

unquestionably the year of the plum. Polly and I found a hedgerow (maybe the one-time boundary of a smallholder's orchard) with every kind of feral variety from full-blown damsons to sloes, including one so round and ripe that picking it was like milking: you simply put your hand under the clusters and they tumbled into it, like a libation. Some had fallen into the stubble, and become impaled on the straws, like extravagant fruit lollies.

Domesticated plums originated with the humble European sloe, crossed with myrobalans (cherry-plums) from the Middle East. Hence Damascenes, Damasks, damsons. Seventeenth-century plum varieties are like fruits from the Song of Solomon: the Great Damask Violet, the Fotheringham, the Perdigron, the Cloth of Gold. I hoped ours might be John Evelyn's favourite, the Dark Primordial. Its frost-at-dawn bloom and pristine eggishness made me want to put one in an egg-cup and eat it with a spoon. But we made our own wild prunes instead, and a dark jam from a recipe by Gisele Tronche. She mixes the damsons with other hedgerow fruit and with her magic ingredient, cumin seed, and describes the result as a '*humeur noir*', with 'the colour of a good, healthy, black-tempered funk'. The taste was like that, too, primeval and feisty. John Evelyn might have liked it himself. The seventeenth-century diarist and royalist supporter, who'd argued for the afforestation of Breckland, was also, surprisingly, a passionate enthusiast for fruit and vegetables. His impish *A Discourse on Salletts* reveals him as an evangelical vegetarian and animal-rights supporter. In a spectacular re-working of Genesis, he attributes the Fall not to picking the fruit of the tree, but to *not* picking fruit: 'The Hortulan Provision of the Golden Age fitted all Places, Times and Persons; and when Man is restored to that State again, it will be as it was in the beginning.'

My bread-making began to go distinctly wild, too. I'd been making my own since the spring, inspired by Polly's weekly ritual. I loved watching her knead, that wiriness and rhythm and finger-tip instinct. It was, of course, undeniably sexy, but also had the inherent attractiveness of a thing done well. I wanted to try and do it myself, and peered and listened, and poked my fingers in the dough to try and absorb what it should feel like. And I read Eliza Acton, too. Pour in the warm water, she wrote in 1857, 'and gently mix in the flour surrounding the yeasty part, and go on doing this, sides to middle, with a happy pawing action, like a contented cat, till the dough is all one even mass of warm soft resilience'. Now *that* I understood. It was an action I'd been familiar with since I'd first had a cat-net put over my pram.

Gradually, I began to get a feel for it, and to ignore the instructions on the back of the yeast packets. In what began as a feeble hypochondriacal gesture, I began trying to make non-wheat loaves. Some were a disaster. Buckwheat dough, even with yeast, was like mud to knead. Without yeast, it baked to the scent and consistency of an old fungus. I dabbled with millet and maize, with pure oat loaves, and with mixtures of all three. Breads of strange colours and unaccustomed textures materialised. But none had the stoneground chunkiness of wheat-bread. It was when I turned to adding nut flours that things happened. I began with reconstituted chestnuts, then almonds and hazelnuts, ground up in a blender. Then I struck gold with a mixture of ground pecans and wheat flour. Somehow the oil from the nuts had coated the outside of the loaf, and generated a crust like a fragrant biscuit. It is now my speciality bread, made on feast-days. But next year I am going back to the neolithic, and trying the weed-seeds that the first villagers made into unleavened bricks to see them through the winter.

*

I'd dreamed of going to America since I was a teenager. I'm not quite sure why, except that half a lifetime of listening to its music and watching road-movies and reading about the western deserts had given even the United States' grossest elements a perversely glamorous image. A bit of me had always wanted to sit in a diner, and ride in a yellow cab, as if they were moments from a favourite fairy-story. But I'd never made it. An inexplicable terror of long-distance travel, of being 'out of my knowledge', had made me back out of two or three chances. It was a legacy of that timorous, highly strung childhood that I'd never quite grown out of. Well, now I *had* grown out of it, and getting to America began to shape up as a dare on myself, a final test of fledging. And I had an excuse to bolster my courage. I wanted to glimpse wildness taken to the limit, experience some scrap of real wilderness. And maybe understand a little more how, for all its aggressive politics, America has never entirely swallowed the British belief that every last inch of the land must be managed. Nature is taken *seriously* there. Annie Dillard and Gary Snyder had won Pulitzer prizes, for goodness' sake, and I wanted to touch their motherlode.

In New York Polly and I stayed in the Algonquin and ate food from all five continents. We spent Sunday in the cheerful mêlée of Central Park. Polly went skating, and I saw a great horned owl slip through the trees above the frisbee players. When we left town, travelling south on the train, we glimpsed a whole urban savannah between the railway and the condominiums: acres of reed-swamp lit up by the scarlet foliage of sumach scrub, and snowy egrets hunched under the stanchions of the Newark dockyards. It suddenly all seemed oddly familiar, and not at all out of my knowledge.

We were bound for the Chesapeake Bay. Polly, who'd spent her babyhood in New Jersey, had old friends to see, and I had a lecture to give. Our host's son was a farmer, growing soya-beans for conversion to gas on half his land, and allowing the rest to evolve into prairie, under a federal conservation programme. We

walked over the farms. They kept reminding me of East Anglia –
flat, damp, slightly wooded, and with a very well-understood
nature. We were initiated into the dangers of poison ivy, given
sassafras roots to smell, and saw monarch butterflies – like flecks of
amber stained glass – flitting over the bean-haulm. A chevron of
birds, black and white wings against a pure blue sky, flew high
overhead. I couldn't make out what they were at first, and, thinking
of home, wondered if they were cranes. But it dawned on me that
they were snow geese, on the half-way point of their great
migration south. They were still calling after three thousand miles.

We stayed for a while at a Maryland Bed and Breakfast, a
three-storeyed colonial house restored to an apotheosis of what it
might once have been, or dreamed of being. It had hand-painted
wallpaper. The ceiling cornices were modelled on a recognisable bit
of the Alhambra's ceiling. The rooms were weighed down with the
kind of furniture you might find on a river-boat. It was pure
American Gothic, and it was no real surprise when we woke up to
see a turkey-vulture's wing draped over the bedroom window.
Outside the birds were everywhere, hanging like scarecrows in the
ornamental conifers, jostling for throw-outs on the lawns. Our
landlady, the restorer, wrung her hands when we mentioned them.
'They always come around Hallowe'en time. They give the wrong
impression of the place. My husband throws tennis balls at them to
get them off the roof.' Apparently they eat insulation and leave
unsavoury odours. I thought they were the wildest bit of America
we'd seen.

But if we wanted to see real wilderness we knew we would,
ironically, have to drive. I'd found a place on the map, mid-way
between Norfolk and Suffolk (VA), called the Great Dismal
Swamp, and given where we'd come from, this seemed too good a
joke to miss. So we hired a car and took off south. It was a culture
shock for me. I hadn't taken in the scale of the distances, or the

extent to which the American road is a self-contained, self-sufficient habitat. You are not expected to leave it. Little detours to find a place to stroll, or sleep, were fruitless. The roadsides were either built-up or fenced off, and the villages in black-out by six in the evening. The long drive south to the swamp looked increasingly quixotic, given that we'd only got a few days left. We cut our losses and motored west, down Route 66 towards the Appalachians.

It was Hallowe'en when we reached the foothills, and around the Baptist chapels and the gardens with their Bambi statuettes was an extraordinary display of mock necrophilia: off-the-peg skeletons, customised tombstones, ghosts made out of polythene bags and whole houses lit up like sorcerers' caves. It was also the start of a real southern heatwave. Everybody was out on their verandahs. In Harper's Ferry, Jason, antique supremo, was sprawled soporifically across several rocking chairs, but inside his shops, Shenandoah rustic pottery was marked up at thousands of dollars a piece. At the yard-sale next door, a true cousin to the Greenpeace Fair, they were hawking liberated Gideon Bibles for dimes. We kept to the back-country roads where we could, and crossed the Shenandoah River at dusk over a bridge no more than a foot above water. A young couple were washing car tyres in the river under flocks of bats. In Pleasant Valley, the men were building up their woodstacks, and Dane, of 'Mountin' Man Taxidermy', had his new four-colour brochure in the store, next to the jars of alfalfa seeds. 'Deer mounts include banded competition eyes. Add $75 for an open mouth.' But we could not find our way into their patch of the woods. The smallest country roads were lined with houses – except where there were 'Posted Private' hunting notices pinned to every roadside tree. Even deep into the state forests, there were long rows of clapboard summer cottages, with their own painted mail-boxes and satellite dishes. The road began to feel like the new frontier, the place where you stake your claims, peddle your wares, fly the flag, do anything but let other people across.

In the evenings I browsed in books on the American wilderness, and on the curious and ambivalent role it plays in the national culture. It is both a symbol of a free-born nation, something to be treasured, and a challenge to the frontier spirit to 'reclaim'. It is loved, lusted after and agonised about. To those of us from Britain, who have very few places left that could even remotely qualify as wilderness, it is the quibbling about definition that is most perplexing. What does wilderness mean? A place unchanged by humans, or unvisited by them? Or just something more subtly undetermined by us? In the purest terms, of course, there is now nowhere on the planet entirely unaffected by human activity: global warming and the ubiquitous spread of toxic chemicals in the seas and atmosphere have seen to that. There are those, too, who resist the concept for political and cultural reasons. It is seen as socially exclusive, a new form of colonialism. It appropriates for the distant rich the living and working territories of marginalised peoples. It is a discriminatory category, demoting the value of more sullied places. Even the very word is regarded, in some deep ecological circles, as self-contradictory. The moment a wild place is recognised, named, mapped, it is, in that very act, domesticated. In his study *Wilderness and the American Mind*, Roderick Nash argued that 'wilderness was a state of mind – a perceived rather than an actual condition of the environment'. One child he spoke to saw wilderness as 'the dark space under my bed'. For Wordsworth it was a condition of the spirit as much as the land. His much-quoted phrase 'a wilderness is rich with liberty' occurs in a poem about the release of two goldfish into the liberating wastes of a Lake District pond.

For Thoreau, too, wilderness was a nebulous idea, rather

than an actual tract of land. His experience on Mount Ktaadn was a one-off, a near-religious encounter you sense he did not want to repeat. In his journals, and in *Walden* especially, 'the wilderness' is either just the general wild country round about (especially, the Massachusetts swamps) or, in a way that prefigures Colette's vision, a place to dream about, not 'be'. He thought that 'village life would stagnate' if it did not have occasional access to the 'tonic' of ordinary wild places nearby; and that it was essential – but enough – just to know of the existence of the inaccessible, 'the unexplorable . . . the unsurveyed and unfathomed'. 'We need to witness our own limits transgressed, and some life pasturing freely where we never wander.' In his late book *Wild Fruits* he toys with the notion of what one can only call civic wilderness: 'I think that every town should have a park, or rather a primitive forest, of five hundred or a thousand acres, either in one body or several, where a stick would not be cut for fuel, not for the navy, not to make wagons, but stand and decay for higher uses – a common possession for instruction and recreation.'

The Shenandoah National Park, 280 square miles of forest in the Blue Ridge Mountains, is close in spirit to Thoreau's idea, and is presented as a great showcase for the 'instruction and recreation' of the eastern states. I asked a guide at the visitors' centre where Annie Dillard's Tinker Creek was, and he pointed out a remote spot in the south-west corner. It looked about as far away as New York, but added a little *frisson*. We walked out along one of the waymarked trails, well aware that what we should have been doing was packing a tent or a canoe, stripping off all our cultural baggage, and trekking off into the west. But it was an enjoyable woodland stroll, and, as on our farm walk, we saw new things: chipmunks with flag-pole tails, osage-orange fruits as big as tennis balls, hickories and maples and American beech, all putting on their fall show.

We browsed like this on the edge of wildness all the way back to New York. A bright scarlet cardinal flew in front of the windscreen. We saw twenty dead racoons to one live – all quite whole, as if they'd just been tossed off the back of cars, not run down by them. We strolled along the boardwalks (those roads again) of another cluster of wildlife refuges, and got ourselves evicted from a National Wildlife Refuge near Washington for being there after closing time. We knew the wilderness was out there, but it remained elusive. The reasons were all on our side. I hadn't done my preparation. We had too little time, too few maps, and absolutely no equipment.

But I began to wonder – I hope not just rationalising my own naivety – if wilderness was really what I wanted, or *should* want. Truly wild places should be for the wild creatures that live there, and only secondarily to give us revelatory experiences. If we go into them it should be as a privilege, and on the same terms as the creatures that live there, unarmed and on foot. They cannot be treated as convenience habitats, available off-the-peg, and in that sense America has got it right. But what I missed was some common ground between the wilderness and the thoroughly domesticated, some accessible country – real and metaphorical – beyond the boardwalks and the forest condominiums and the hunting reserves. I realised that what touched me most was not wilderness as a special, defined place, but the quality of *wildness*, Dylan Thomas's 'force that through the green fuse drives the flower', the untidy, energising edge of all living systems. True wildernesses must be defended at all costs, for the sake of their rightful inhabitants. But I felt that I could settle, like Thoreau and Colette, for just knowing they were there, and leave the real experience to my imagination. Our biggest challenge as a species is to work out a common arena with nature, a hinterland where we can accept each other's company, and live out a relationship somewhere between the ten-day wilderness experience and the short stroll along a fenced trail. I thought about those spontaneous swamps by the Newark railroad, about the turkey-vultures outside our window, and wondered if they weren't just as invigoratingly wild as anything we might have seen in the Great Dismal Swamp.

The odd — and heartening — thing is that many of America's special reserves aren't ancient wild places, but restored ones. There was virtually no woodland at all in the Shenandoah Park when it was dedicated in 1936, on an area of depopulated dirt farming. Now two-fifths of it is an official US wilderness. All through the eastern states new-growth forest and the creatures that go with it are reclaiming abandoned agricultural land. Vermont has gone from 35 per cent woodland in 1850 to 80 per cent today. Martha's Vineyard is oak forest again. Coyotes are back in Cape Cod, and moose have started playing Russian roulette on Route 128, America's 'computer highway'. The wild is creeping back along the new frontier.

*

Back home, the autumn colours were putting Virginia to shame. The field maples and hornbeams blazed in the hedges. The rows of highway authority dogwoods shone like strings of garnets. In the honeyed afternoon light the whole landscape seemed caught in amber. 'Suppose', Loren Eiseley once wrote, 'we saw ourselves burning like maples in a golden autumn . . .', that we could 'disintegrate like autumn leaves fret away, dropping their substance like chlorophyll, would not our attitude toward death be different?'

The fens took Eiseley's cosmic view of change, and declined to be caught in amber or any other kind of preservative.

Snubbing all those who believe in static states, and making sport with my own vision of East Anglia's primal wetness, they simply dropped their substance and turned bone-dry. The meres evaporated and our few snipe fled to the coast. The raft-spider pools sank to nothingness, leaving forlorn and wispy sockets, and no one was quite sure where their inhabitants had gone. There was a curious feeling in the air, a slightly gleeful sense of playing truant from the proper order of things. When, we wondered, would normal service be resumed? When would we have to be glum again?

On my regular sorties out into the valley, checking out the lie of the land, I got the distinct feeling that something was up. The summer had changed England, but America seemed to have subtly changed me, too. The very ordinary wildness that, as greenhorns, we'd enjoyed there, made me look at our own tamed acres with a new respect, and a willingness to suspend judgement about their domesticity.

So, for a change one day, I turn north from the house, away from the valley fens and up to the edge of Norfolk's grain plateau. They call this 'England's Bread Basket'. It is landscape taken to its economic limit. The fields stretch from lane edge to horizon, dwarfing the hedges and copses into irrelevant doodles. The grain silos overtop the churches. The sugar-beet harvest has started early, and the huge machines are chopping through the fields like galleons, trailing wakes of gulls. For a few weeks weeds and stubble will be the field-crops. The sharp straw seems absorbent, as full of mirages as a heat haze. Birds materialise from seemingly blank spaces. A flock of goldfinches erupts about 20 yards in front of me, a puff of sequins and chaff. I first see the sparrowhawk as not much more than a shadowy thickening of the air. It's a female, fallow-brown, gliding impossibly slowly only a yard or so above the furrows. Her wings have the slightest upward tilt, like a harrier.

Her head is bowed, glaring, implacable. It is as if she is drawing a scalpel across the field, opening its skin for the slightest tremor of life beneath. The poet Kathleen Jamie once likened the bond between a male and female peregrine, standing at a distance from each other close to the nest, to the electrifying link between two flamenco dancers, known as the *duende*. That is the magnetic pull this sparrowhawk has with the field. She draws it into tension, into attention. The wagtails' nerve breaks first. From nowhere half a dozen birds spring up and pursue the hawk. They move like the waving tail behind a Chinese dragon. A cloud of larks erupts ahead of it. Suddenly the hawk has had enough and, in a flourish, seems to unfold. It changes shape, turns bulky and furious, and careers away to the east.

Walking up here today is like gazing into a crystal ball. Echoes of ancient landscapes surface from the flats like overlooked birds. I pass Nordle Corner, Folly Lane, pointless dog-legs, inscrutable mementoes of some older order in the fields. Then, just before dusk in Hall Lane, I see half a dozen birds darting low over the banks of the Hall's old park. The golden plover – Norfolk's 'whistling plover' – are back. About fifty birds are feeding with lapwings in the winter wheat. Through my glasses they look as tranquil as doves. But in flight they're wild and tremulous. They scramble repeatedly into dark and reckless packs, as if methodical feeding were just too mundane for birds that have flown here all the way from the tundra.

It was a revelation, and I began going up to the fields most days at last light. For a few weeks before the new wheat grew too tall there were birds everywhere. I saw more sparrowhawks working the stubbles – sometimes two at once – and immense flocks of fieldfares, every bird pointing and edging in the same direction, heads held high, chests out, onward, onward. But the plovers were more elusive. Mostly they were in the distance, or high above me, slicing through the loping lapwings like arrow showers. A tiny bit of me wished they weren't here, making excuses for these spendthrift arable wastes. But that is the benediction of the wild, to see opportunity in the briefest of openings, the narrowest of windows. Tomorrow the fields would be clotted with overblown cash-crops. Today they were dancing.

*

It was almost the first anniversary of my arrival in the valley when I heard about the barn owl. The word came from a friend, who'd been tipped off by her window cleaner, who regarded it as a matter of course: yes, he saw a white owl most evenings when he was taking the dog for a walk, in that little side valley, back of Botesdale.

I went there at dusk the next day. I settled down close to where I guessed it might hunt, in a stretch of streamside rough, scattered with newly planted saplings. I must, I think, have been very still. A roe deer gazed at me from the other side of a hedge. A woodcock slipped into the stream just a few yards in front of me, preening and probing. In the half-light its two pale back-stripes seemed to sway disembodiedly against the water, like eels. At forty minutes after sunset the barn owl just materialised, rose up out of the grass. I guess it had been there all the while, gobbling mice, hunched-up on the ground. It took off, soft as thistledown, its head like a quite separate creature riding shotgun out front. It passed through the saplings, wing-beats quickening to negotiate the gaps. It was winnowing the grass, threshing it for food. I could see the last light from the west shining through its wings, marking out the dense primaries from the almost translucent trailing feathers. It seemed at that moment to have four wings, two in the day and two in the night. They were shuttling the twilight forward. Then it stalled, hovered on tip-tail, and plunged into invisibility again.

As I walked home, the twigs seemed to be encased in the faintest golden aura, and the air full of the flickering shadows of birds that I had never dreamed would be there. Evening was when the action happened. When the plovers had vanished the great dramas of roosting began. Every late afternoon immense flocks of rooks and jackdaws poured towards a roost-site I hadn't yet discovered. They put down here and there on the way, and turned the fields black. Over at a mere about a dozen miles to the west, I watched the evening wildfowl fly-around. Bands of wigeon and shoveler hurtled about the sky, beating the bounds of the water and then shooting off on tangents over the nearby houses and fields. It was an exultant, infectious display. Other birds got caught up in it. I spotted a single starling flying deep in a wigeon flock, then a dozen teal riding with the mallards and matching every flick and swerve.

We went back one evening to the place where we'd first seen the cranes a year before. It was warm and balmy, and we followed our old route through the scrub, past the boathouses, and round the back of the mere. There were greylags in the damp meadows, and a few snipe. Then a couple of hundred yards to our right, five huge grey birds appeared above the reeds. The cranes were coming in to roost, flying low, right past the windows of the shore-line bungalows. They were in single file, a line which rose and fell like the swell of the sea. They were close enough for us to see their red caps and black throats. They hugged the lea of the sand-dunes till they were out of sight. We followed them, out as far as a derelict mill, its arms stark against the setting sun. There was no sign of them, and, for a while, the place seemed emptied of birds.

We turned for home, and in that moment of turning, two

barn owls flew out of the mill, wings held still and close, as if they had been lobbed. They began quartering the marsh, two exquisitely different birds, one pale honey, the other darkly chequered with chestnut, like a piece of marquetry. And as if this were the signal that the rites of dusk had begun, the sky began to fill. Marsh harriers, playing with the wind as ever, glided to their roost-site in the marsh scrub. Ten more cranes skimmed briefly across the reeds, bound for some hidden lagoon. For an extraordinary moment I had, in a single field of view, two cranes heading directly towards me, three hen harriers, sharp-winged and long-tailed, and, in the far background, thousands of pink-feet geese flying down to a field.

It was a hallucinatory sight: a vision of quintessential East Anglia, with its windmill towers rearing out of the reeds into a vast, darkening sky, but with such a multitude of birds that we could have been on the African plains. It was cosmopolitan in another way, a gathering that connected the deeply local with the great currents of life that flow across the planet. The cranes, by right, should never have stayed here in 1979, but flown on to Spain. The marsh harriers have taken to spending the whole year here, instead of migrating south. The pink-feet come from Iceland and Greenland, and will return there in the spring. The hen harriers come from northern Europe and Holland, and will go back, too. But this was the site of the very last pair to breed in Norfolk, and I dream, that like the cranes, one year they may stay.

What do these vespers rituals mean? Conventional theories explain communal roosting very plausibly in terms of 'safety in numbers' and the sharing of information about food sources. But as so often in nature, what happens is too extravagant, too excessive, to be so simply utilitarian. The huge assemblies, the prolonged and festive flight displays, the mixing of species, all suggest that something else is going on. Is it too anthropomorphic to imagine that, like the red kites, other birds like to see in the night in company, to put on a show for mutual reassurance against the dark? It's a shared moment, that familiar glimpse of another traveller on the same road, unknown, probably unknowable, but bound home too.

I'm becoming crepuscular myself, a born-again dusker. Back in the spring, still in the last throes of my long low mood and unsure of my ground, I saw the dusk as a medium for confronting the sombre realities of the place I'd fetched up in. I half wanted to be disillusioned, snubbed. And I saw, in consequence, just the bare bones of the landscape. Now that image has flipped over, come out like a positive print. I know the barrenness of what is there, but in the half-light it retreats into the background, and I notice instead the brilliant fringes of things: tricks of the light through the tracery of twigs; the swellings of the ground that will in just five months be exuberant vegetation again; the sparrowhawk, and not the vacuum it seems to be drawn to, and the barn owl I thought had abandoned us. I suppose a therapist would say I had reconfigured the dark.

*

Is this what Thoreau was getting at in the phrase for which he is most remembered – 'In Wildness is the preservation of the World'? That the unmanaged energy of nature is the source of both the world's stability and its new beginnings? Of, metaphorically, the ancient wood and the quicksilver wet? The phrase occurs almost as a throwaway remark, in a short and maverick book called *Walking and the Wild*, in which he floats his idea of the 'westward impulse' of life.

> I derive more of my subsistence from the swamps which surround my native town than from the cultivated gardens

in the village . . . When I would recreate myself, I seek the darkest wood, the thickest and most interminable, and, to the citizen, most dismal swamp. I enter a swamp as a sacred place, – a *sanctum sanctorum*. There is the strength, the marrow of nature.

But Thoreau is being a mite disingenuous here. In his retreat at Walden Pond he was essentially a commoner, a literary peasant. He measured the depth of the pond. He fed the mice that came into his workroom. His joyous account of working in his bean-field, barefoot and revelling in the birds, is a classic piece of contented pastoral. Walden Pond itself was far from being a wilderness. It was fished, had its ice harvested in winter, and was surrounded by worked woods and farms. And yet Thoreau saw its essential wildness, the 'tangled fringes'.

Later in *Walden* he writes an account of the thawing of a frozen sandbank on a railroad cutting in the village that has become one of the seminal texts of cultural ecology. He'd glimpsed something vegetative about the patterns the trickling sand made, and launches on an audacious piece of writing which, in a tumble of visual puns and risky analogies, links the flow of the sand with the growth of plants, and eventually with language itself. 'As it flows it takes the form of sappy leaves or vines, making heaps of pulpy sprays a foot or more in depth, and resembling, as you look down on them, the laciniate lobes and imbricated thalluses of some lichens.' Then he's reminded of birds' feet, brains, excrement. The pattern is imitated in bronze, 'a sort of architectural foliage more ancient and typical than acanthus'. Then the whole 'lava' flow resembles the inside of a cave, until it flattens at the foot of the slope, into 'banks, like those formed off the mouths of rivers'. He feels he is in the 'laboratory of the Artist who made the world and me, – had come to where he was still at work, sporting on this

bank, and with excess of energy strewing his fresh designs about'.
He sees in this sandy overflow

> an anticipation of the vegetable leaf ... *Internally* whether
> in the globe or animal body, it is a moist thick *lobe,* a word
> especially applicable to the liver and lungs and the *leave*s of
> fat (*labor, lapsus,* to flow or slip downward, a lapsing;
> globus, lobe, globe; also lap, flap, and many other words)
> *externally* a dry thin *leaf* even as the *f* and *v* are a pressed
> and dried *b*. The radicals of lobe are *lb*, the soft mass of the
> *b* (single lobed, or B, double lobed) with a liquid *l* behind it
> pressing it forward. In globe, *glb*, the guttural *g* adds to the
> meaning the capacity of the throat. The feathers and wings
> of birds are still drier and thinner leaves. Thus, also, you
> pass from the lumpish grub in the earth to the airy and
> fluttering butterfly.

And so his fantastic, elaborate vision continues during the next
morning's thaw, through re-imaginings of ice crystals and blood-
vessels and the leaf-like shapes of hands and ears. 'Thus it seemed
that this one hillside illustrated the principle of all the operations of
Nature. The Maker of this earth but patented a leaf. What
Champollion will decipher this hieroglyphic for us, that we may
turn over a new leaf at last?' Thoreau had produced his own comic
creation myth, which, in a way that would satisfy any modern
literary critic, is as 'true' to the external world as it is to the drive of
the text, and to the extraordinary willow-leaf parabola of his
extended metaphor. (Ruskin would have loved this passage, if only
he'd had the sense of humour to see the joke.) It is also a plausible
attempt to marry a vision in which pattern in nature coexists with
wild spontaneity. But the final comic turn is that the wildest thing

here is Thoreau's own imagination, produced by a man who, as he writes, is 'but a mass of thawing clay'.

*

Now the rains have started again, I can see the valley's own sand runnels forming as they did last autumn, trickling out of the ditches where the drainage grups have been cleared. But I notice the water itself more than the earth it drives. Again it's following ancient patterns, filling dips in the greens that were invisible in summer and deeper hollows that may have been here since the Ice Age. Yet the shapes it makes are amorphous, unbiddable. Their edges and direction are unpredictable. They're fuelled by invisible and moody springs. Already they're making their own luck, moving on from the beachheads created by last autumn's rains and the summer's heatwave.

Edward O. Wilson, the great biologist who gave the world the concept of biophilia, hopes that one day all such processes will be understood and made accountable by science. In his book *Consilience* he dreams of a grand unifying science that will reveal, explain and predict all things under the sun, from the movement of water to the imaginative processes of artists. It is the same questionable project which Francis Bacon initiated four centuries ago, determined on 'enlarging of the bounds of human empire', and it is a perplexing dream from a man who has so eloquently named and celebrated the idea of 'biodiversity', as the sum of all the teeming organisms and their interrelations in an ecosystem. To wish to contain and know that wild, proliferating edge is to wish to stop nature in its tracks, to put it in a cultural reserve. But it's an unlikely end. In the nature of things life will always keep one step ahead of the measurers and managers. A ceaseless, gratuitous, inventive bodging is what keeps the world going. Wouldn't it have been simpler, Annie Dillard enquires of God, 'just to rough in a slab of chemicals, a green acre of goo? . . . The lone ping into being of the first hydrogen atom *ex nihilo* was so unthinkable, violently radical, that surely it ought to have been enough, more than enough. But look what happens. You open the door and all heaven and hell break loose.'

Lewis Thomas called DNA's habit of suddenly mutating, out of the blue, 'the wonderful mistake'. You can see an equally wonderful refusal to stick to the point as a characteristic of all living matter. In the inexplicable intricacy of a redcurrant flower, in the merging and drifting clouds of orchid varieties on the fen, in the milling festivals of swifts and the dances of cranes, in the nightjar's all-enveloping song and the house martin's improvisations, there is what Annie Dillard called the world's 'free, fringed tangle. Freedom is the world's water and weather, the world's nourishment freely given, its soil and sap.' It is also, of course, the world's pain, the wrecking gales, the collapsing membranes, the awareness of death that maybe makes us yearn so much to leave our mark on the earth.

The idea of a 'nature cure' goes back as far as written history. If you expose yourself to the healing currents of the outdoors, the theory goes, your ill-health will be rinsed away. The Romans had a saying, '*solvitur ambulando*', which means, roughly, 'you can work it out by walking', including your own emotional tangles. The medievals made mass pilgrimages to rustic shrines. John Keats, mortally ill with tuberculosis, fled to the Mediterranean to find that 'beaker full of the warm South', away from that place 'where youth grows pale, and spectre-thin, and dies'. 'The country, by the gentleness and variety of its landscapes,' wrote the philosopher Michel Foucault, 'wins melancholics from their single obsession "by taking them away from the places that might revive the memory of their sufferings".' My friend Ronald Blythe, the great chronicler of East Anglian life, has shown me the sites of the local sanatoria, where impoverished twentieth-century TB patients were sent. He'd seen them lying outside in pram-beds in all weathers, sometimes with snow covering the mackintosh aprons which kept their blankets dry (nature of course being anything but uncomplicatedly full of warmth and gentleness).

The idea was to submit to nature, to hope that it would 'take you out of yourself', dissolve the membrane between you and the world of health to which you naturally belonged. But that was not how it happened with me. I'd tried repeatedly to exorcise my depression by this kind of exposure, but I was too disconnected by then, and all I felt was a kind of rebuke, a clear statement that I was no longer part of that world. And a sense of shame that it was wasted on me, in some sense spoilt by my lack of engagement.

What healed me, I think, was almost the exact opposite process, a sense of being taken not out of myself but back *in*, of nature entering me, firing up the wild bits of my imagination. If there was a single moment when I was 'cured' it was that flash of loving inspiration by Polly, that sat me down under the beech tree in my old home, and made me pick up a pen again. It was those first stumbling imaginative acts that reconnected me, more than the autumn breeze through the trees. The physical rejoining came later, and my translation from the depths of forest country to the bright and shifting landscape of the fens was a huge metaphorical support. I really did have to listen, and look up. And somehow, in this process of out-going, I've found a new confidence. I'm not timid or phobic any more. And I seem to have become garrulous and inquisitive, too, the kind of person who drags you into conversation in a shopping queue and interrupts your meditations in the fen.

But in all honesty, I'm not sure I am confident enough to say I'm 'cured'. I doubt that I will ever go down to the depths again, but

I'm still something of a wind-harp, over-strung and susceptible to emotionally stormy times. Polly generously tries to persuade me that this is all part of a general sensitivity, on a par with my swift-withdrawal syndrome. But I fear that is taking holism a tad too far. What I try to do (not very successfully) is not resent these quirks in myself, to accept my chances in 'the mill of the world's breathing', even if it does sometimes blow me about a bit.

And I've become oddly attracted to the idea of 'vegetative-ness', to trying to tune in with other kinds of mind that operate on the earth without the privilege of self-consciousness: they're part of the commons, too. Not vegetative retreat, of course; that would be negativity even in a vegetable. But to learn a little more about 'vegetative advance', to work at gently joining our ancient, shared senses more closely with our actions – that might be no bad ambition for our adrenalin-soaked culture. We badly need to find ways of juggling that simultaneous existence in our own world and in nature's.

To Blackie such crossing-over is second nature. I've come down to the end of the garden to call her. She spends a lot of time now a long way from her home, maybe playing with the feral cats on the Ling. I spot her more than 200 yards away, bounding back towards the house. She skirts hedges, leaps over tussocks, making straight for me. Then, when she's maybe 30 yards off, her approach changes completely. She switches to a walk, making single paces to the left and right, sniffing a few plants, not catching my eye. What is she trying to say? That she is confident that I will wait for her, and not run away? That she is still an independent animal, and is coming of her own free will, not out of habit or bidding? I am putty in Blackie's paws, and mustn't let my imagination run away with me. She and her kind are not wild animals, and have no real role in the negotiation we have to work out with nature. Yet cats seem to me to be messengers. Their effortless passing between the wild and domestic worlds suggests the kind of grace we need as a species to move between nature and culture. I think of some lines from Christopher Smart's extraordinary ecological hymn, *Jubilate Agno*. It's the passage usually known as 'To my cat Jeoffry', but it might be a dream for our species, too:

> For he can tread to all the measures upon the music.
> For he can swim for life.
> For he can creep.

What Blackie does not know, and would not care if she did, is that I have come to say *au revoir*. Polly and I have found a house of our own, and I am moving again.

*

But not far this time, just half a mile north over the river, and up to the 45-metre contour line, a mountain climb by East Anglian standards. This was done with a bit more forethought than my migration to East Anglia, but still seems a change conjured up by fortune and serendipity. But this time, I'm not a carefree lodger but a co-owner. The basic survival skills I needed for the house on the Ling won't suffice. I'll have to learn to negotiate the borderlines between nature and culture in my own life, too. And as if to emphasise this, the land that goes with the house looks like an arena which will make my theoretical musings about the wild and the tame into a very practical debate.

Both sides seem already to have staked their claim. The house was a small farm, built in the early 1600s from timbers that in places still carry fragments of impacted bark. On almost all of them there are traces of some inexplicable snake-like markings. The walls are made (home-made, probably) from clay and flint and brick, and above them some of the roof-laths are lashed down with woven willow.

Our patch is surrounded on two and half sides by Norfolk's arable plateau, but more ancient farming systems have left their echoes. There's a pond that was probably first dug to provide clay for the walls, and then used as a 'retting' pool, for soaking hemp fibres clear of their husks. The Waveney Valley's ancestral crop was grown, till 150 years ago, virtually in our back garden. The 1839 Tithe Map shows 10 acres of 'Hempland' attached to the farm, and two rows of orchard trees out the front. It was lived in by two bachelor smallholders. Twenty years earlier, before Enclosure, they would have had rights on the small common which adjoined the house, and whose ghost now survives in a small plantation.

What do you do with a plot as ingrained with history as this? Try and take it back to some imagined earlier state? Back before Enclosure? Back before the house was here at all? The pond lies like some mute oracle at the juncture of the human and non-human worlds, prodigiously deep and blank-faced. The field sparrowhawks use it as the one route along which they can make raids on the garden. A mysterious cold spring bubbles up beneath it. It is, as water always is, a portal through which the wild insinuates itself into the domestic. Already I've seen six species of dragon-fly hawking over it. Woodpeckers drink under cover of the tangled vegetation that drapes its sides, hanging upside down from the tree-roots. But its artificially steep sides, the legacy of its origins, mean that it will always be a resolutely defined garden pond. Should we free it up a bit, scallop in some shallows, give nature more of an edge, trade in some seventeenth-century culture for some twenty-first?

Beyond the pond the decisions may be a little easier. We've about half an acre of grass, which we'd like to tease back into meadow. We've already had it cut. A farmer from Thrandeston came in with a haymaker so big that I blanched at the thought of it coming in the gate and round the herbaceous beds. But he mowed the whole lot in four or five elegant swipes without knocking a single apple off the Bramley. Now we've got a cropped sward speckled with invitingly bare tyre-marks and molehills. What I hope is that these little opportunist patches may become the growth points for a wilder patch, in a simulacrum of the way that grasslands evolve naturally. We'll help them on their way, with seeds gathered from the local greens. It would be good if, over the years, it came to resemble the lost common that once backed on to the house. Or then again, becomes whatever it chooses.

It's late in the year, but Polly has already started planting vegetables. It's the maker in her, and unlike any gardening I've ever seen. She works on all fours, making dibbled rows in circles with her fingers, marking them with stones, hanging up switches of thyme on twigs as insect deterrents, plucking strands of bindweed to tie plants to stakes, hoeing *round* and sometimes transplanting favourite weeds. She's hung up CDs to alert the birds to what's going on (and then has to take them down when they're needed for the computer). At night the discs pulse like fireflies in the light from the house.

It all looks a bit like Wicca to me, but I know that it's also intuitive and playful and learned, more in common with cave-painting than arable farming. This is the cultured end of the garden, but this kind of tacking with the wild seems to me very much in Meeker's comic way, and a style to build on way beyond its borders.

Before long these vegetables will be inside a walled garden, an English *hortus conclusus*. A surveyor did an accurate map for us, and it was no surprise to see that the garden edge, which follows the line of an ancient field boundary, was slightly inclined towards the west. When I measured the tilt, it was, as far as my old school protractor could tell, exactly 4 degrees.

Author's Notes

Page

6. The book Tony Evans and I were working on was *The Flowering of Britain*, first published in 1980.

7. Breckland: The first book on the region, W.G. Clarke's *In Breckland Wilds*, 1925, has never been bettered.

16. James Lovelock, *Gaia: A new look at life on Earth*, 1979. Aldo Leopold, *A Sand County Almanac*, 1949. A landmark book in America, in its pioneering attempt to frame an environmental ethic.

17. The quote from William Fiennes is from *The Snow Geese*, 2002.

18. The film was *Devil Birds* by Derek Bromhall. His book of the same name was published in 1980.

19. Ted Hughes, 'Swifts', *Collected Poems*, 2003. On natural metaphors: Stephen Potter and Laurens Sargent, *Pedigree: Words from Nature*, 1973; Lewis Thomas, various essays on language in *The Lives of Cell: Notes of a Biology Watcher*, 1974, and *The Fragile Species*, 1992.

20. Edward O. Wilson, *Biophilia*, 1984; also Stephen R. Kellert and Edward O. Wilson (eds.) *The Biophilia Hypothesis*, 1993. George Ewart Evans and David Thomson, *The Leaping Hare*, 1972.

22. Ian Sinclair, in his psycho-geographical odyssey round the M25, *London Orbital*, 2002. Muntjac is an Asiatic word, and needs naturalising.

23. Clare. The standard editions are the 9 volumes published by Oxford University Press between 1984 and 2003, edited by Eric Robinson, David Powell and PMS Dawson. Shorter, accessible selections include Geoffrey Summerfield (ed.) *John Clare. Selected Poetry*, 1990. John Clare, *The Midsummer Cushion*, ed. Anne Tibbles, 1978.

28. John Ruskin, *The Eagles Nest*, 1887.

29. Edward O. Wilson, *The Diversity of Life*, 1992.

31. A profound and beautifully written book about cats' minds and behaviour is the psychologist Jeffrey Masson's, *The Nine Emotional Lives of Cats*, 2002.

33. Annie Dillard, *Pilgrim at Tinker Creek*, 1974. Her account of writing this is in *The Writing Life*, 1989.

35. Roy Leverton, *Enjoying Moths*, 2001. Henry David Thoreau, *The Maine Woods*, 1864.

36. Gary Snyder, *The Gary Snyder Reader*, 1999.

37. Jonathan Bate, *The Song of the Earth*, 2000. See also his *Romantic Ecology: Wordsworth and the Environmental Tradition*, 1991.

The concept of 'deep ecology' is still unfamiliar in Britain, despite being a major approach to ecology throughout the rest of the developed world. It is a broad movement, stretching from hard-nosed direct action at one extreme to Buddhist mysticism at another. But fundamentally it seeks an altered mind-set towards the natural world, and a shift away from managerial domination and the assumption of custodianship, towards a more equitable relationship, in which our cultural relations are an intrinsic part. A useful, if solemn, introduction is *Deep Ecology: Living as if Nature Mattered*, edited by Bill Devall and George Sessions, 1985. Major influences include Arne Naess (who invented the phrase), Theodore Roszak, Fritjof Capra, Carolyn Merchant, Paul Shepard, Richard Nelson, Susan Griffin. The writings of Gary Snyder, John Livingstone and Edward Abbey are very accessible introductions to the different perspectives.

'Eco-criticism', of which Bate's *The Song of the Earth* is an example, is the literary wing of deep ecology. See, for instance: Cheryll Glotfelty and Harold Fromm (eds.), *The Ecocritism Reader*, 1996. Karl Kroeber, *Ecological Literary Criticism: Romantic Imagining and the Biology of Mind*, 1994. Robert Pogue Harrison, *Forests: the Shadow of Civilisation*, 1992.

Clare: Eric Robinson and David Powell (eds.), *John Clare by Himself*, 1996. Hugh Haughton, Adam Phillips and Geoffrey Summerfield (eds.), *John Clare in Context*, 1994. Jonathan Bate, 'The Rights of Nature', *The Journal of the John Clare Society*, 1995.

44. Ted Ellis, *The Broads*, 1965. Brian Moss, *The Broads. The People's Wetland*, 2001. William Dutt, *The Norfolk Broads*, 1923. J.M. Lambert, et al, *The Making of the Broads: a Reconsideration of their Origin in the Light of New Evidence*, 1960.

45. For the ecology and mythology of cranes in general see: Peter Matthiesen, *The Birds of Heaven*, 2001.

55. Oliver Sacks, *Migraine: The Evolution of a Common Disorder*, 1970. Alexander and French, *Studies in Psychosomatic Medicine: An Approach to the Causes and Treatment of Vegetative Disturbances*, 1948.

56. For John Clare's illness: Jonathan Bate, *John Clare: A Biography*, 2003. A. Foss and K. Trick, *St. Andrew's Hospital*, 1989. Various essays in *John Clare in Context*, op. cit.

59. Richard Mabey, *Home Country*, 1990.

66. Bernd Heinrich, *Ravens in Winter*, 1990.

73. From The Wildlands Project manifesto. Wildlands is at PO Box 455, Richmond, VT 05477, USA. There are real signs that the idea of big wetland reserves is beginning to take root in Britain, especially in East Anglia.

75. Gary Snyder, *The Practice of the Wild*, 1990.

76. *The Journals of Henry David Thoreau*, 1836–61 (14 vols) edited by Bradford Torrey and Francis H. Allen, 1984. Henry David Thoreau, *Walking and the Wild*, 1851.

79. David Abram, 'Out of the Map, into the Territory: The Earthly Topology of Time' in David Rothenberg (ed.), *Wild Ideas*, 1995. See also his *The Spell of the Sensuous: Perception and Language in a More-than-human World*, 1996.

81. David Dymond, *The Norfolk Landscape*, 1985.

87. Stone Age art: Paul G. Bahn, *Journey Through the Ice Age*, 1997. Jean Clottes, *Return to Chauvet Cave*, 2003. Nancy K. Sandars, *Prehistoric Art in Europe*, 1968. Geoffrey Grigson, *Painted Caves*, 1958. The most audacious new interpretation is in David Lewis-Williams, *The Mind in the Cave*, 2002.

91. Martyn Barber, David Field and Peter Topping, *The Neolithic Flint Mines of England*, 1999. Norman Nicholson, 'Ten-yard Panorama' in Richard Mabey (ed.), *Second Nature*, 1984.

94. Virginia Woolf, *A Passionate Apprentice*, 1990. Margaret Gelling, *Place-names in the Landscape*, 1984.

95. For the hemp industry in the Waveney Valley see: Eric Pursehouse, *Waveney Valley Studies*, 1966; Michael Friend Serpell, *A History of the Lophams*, 1980.

96. *Faden's Map of Norfolk*, 1797. Reprinted by The Larks Press, Dereham, Norfolk, 1989.

98. Roger Deakin, *Waterlog*, 1999.

105. Colette, *Earthly Paradise*, ed. Robert Phelps, 1966.

106. Francis Bacon, *Works*, ed. James Spedding et al, 1870. Carolyn Merchant, *The Death of Nature: Women, Ecology and the Scientific Revolution*, 1980.

108. Lewis Thomas, 'Natural Man', in *The Lives of a Cell*, op. cit. Bill McKibben, *The End of Nature*, 1990.

114. Ian Carter and Gerry Whitlow, *Red Kites in the Chilterns*, 2004.

115. The full story of Hardings Wood is in Richard Mabey, *Home Country*, 1990.

122. Edward H. Whybrow, *The History of Berkhamsted Common*, n.d.

123. Pierre Bourdieu, *Outline of a Theory of Practice*, 1977.

126. Garrett Hardin, 'The Tragedy of the Commons' in Garrett Hardin and John Baden, (eds.) *Managing the Commons*, 1977.

127. E.P. Thompson, 'Custom, Law and Common Right' in *Customs in Common*, 1990. The most invigorating analysis yet of common law and custom. See also J.M. Neeson, *Commoners: common right, enclosure and social change in England, 1700–1820*, 1993. Lord Eversley, *Commons, Forests and Footpaths*, 1910.

132. Henry Reed, 'The Naming of Parts', *A Map of Verona*, 1945.

143. Wayland Wood, just south of Watton, is now a Woodland Trust reserve.

148. John Fowles, 'The Blinded Eye' (1971), in *Wormholes: Essays and Occasional Writings*, 1988. Maria Benjamin, 'To have and to hold' in Kate Selway, *Collectors' Items*, 1996.

150. Margaret Grainger, (ed.) *The Natural History Prose Writings of John Clare*, 1983.

151. Geoffrey Grigson, *The Englishman's Flora*, 1958.

152. William Hazlitt, 'On the Love of the Country', 1814.

157. Auric. Devotees of *Blake's Seven* will get the allusion.

160. Edward Armstrong, *Bird Display and Behaviour*, 1947.

162. Anne Stevenson, 'Swifts', *Collected Poems*, 1996.

166. Richard Mabey, *Gilbert White*, 1986.

175. Edward Armstrong, *The Folklore of Birds*, 1958. Francesca Greenoak, *British Birds. Their Folklore, Names and Literature*, 1999.

176. Claude Levi-Strauss, *The Savage Mind*, 1966, and *Totemism*, 1969. A good introduction to his thinking is *From Honey to Ashes*, 1973. See also Paul Feyerabend, *Farewell to Reason*, 1987, and Mary Midgley, *Science and Poetry*, 2002.

180. Joyce Carol Oates, 'Against Nature' in Daniel Halpern (ed.), *Antaeus on Nature*, 1986.

192. Gilbert White, *Journals*, ed Francesca Greenoak, 1986–89.

197. Lewis Thomas, 'The Music of *This* sphere', in *The Lives of a Cell*, op.cit.
G. Christopher Davies, *The Handbook to the Rivers and Broads of Norfolk and Suffolk*, 1882. Henry Doughty, *Summer in Broadland, or Gipsying in East Anglian Waters*, 1889.

199. Joseph Meeker's seminal book, *The Comedy of Survival*, evolved through three sub-title changes: *Studies in Literary Ecology* (1974), *In Search of an Environmental Ethic* (1980) and *Literary Ecology and a Play Ethic* (1997).

205. John Evelyn, *Acetaria: A Discourse on Salletts*, 1699, and *Compleat Gard'ner*, 1693.

210. Roderick Nash, *Wilderness and the American Mind*, 1967. Max Oelschlaeger, *The Idea of Wilderness*, 1991. David Rothenberg (ed.), *Wild Ideas*, op.cit.

211. Henry David Thoreau, *Walden*, 1854. *Wild Fruits*, ed. Bradley P. Dean, 2000.

213. For an account of woodland regeneration on the USA's eastern strip see Bill McKibben, *Hope, Human and Wild*, 1995. Kenneth Heuer, *The Lost Notebooks of Loren Eiseley*, 1987.

222. Edward O. Wilson, *Consilience. The Unity of Knowledge*, 1998.

223. Annie Dillard, *Pilgrim at Tinker Creek*, op.cit.

226. Christopher Smart, 'Jubilate Agno' in Karina Williamson and Marcus Walsh, *Christopher Smart: Selected Poems*, 1990.